Travel and Lifestyle

This engaging book presents nine empirical chapters that explore topics such as lifestyle entrepreneurship, lifestyle mobility, luxury experiences, and tourism-related well-being.

Unlike most research focusing on Western contexts, several of the studies involve Asian regions (particularly China, including Hong Kong and Taiwan) and capture the growing popularity of Asian perspectives. This edited volume, authored by researchers across China, New Zealand, the US, the UK, and Portugal, provides researchers and practitioners in tourism and hospitality, along with readers interested in the general "travel and lifestyle" domain, timely and relevant knowledge. The editors hope that these carefully chosen chapters will inspire future studies and will give its readers a fresh insight in lifestyle's role in tourism.

The chapters in this book were originally published as a special issue of the *Journal of Travel & Tourism Marketing*.

Mimi Li is Associate Professor in the School of Hotel and Tourism Management at The Hong Kong Polytechnic University. Dr. Li is specialized in tourism marketing, tourist behavior, tourism planning, and China tourism related issues.

Xiaoting Huang is Professor in the Department of Tourism Management at Shandong University. She is specialized in tourists' space-time behavior, tourism planning, and travel mobile behavior issues.

Han Shen is Professor in the Tourism Department at Fudan University, Shanghai, China. She is specialized in consumer behavior, city branding, destination marketing, and advertising.

Travel and Lifestyle

Edited by
Mimi Li, Xiaoting Huang and Han Shen

Routledge
Taylor & Francis Group
LONDON AND NEW YORK

First published 2022
by Routledge
2 Park Square, Milton Park, Abingdon, Oxon, OX14 4RN

and by Routledge
605 Third Avenue, New York, NY 10158

Routledge is an imprint of the Taylor & Francis Group, an informa business

British Library Cataloguing-in-Publication Data
A catalogue record for this book is available from the British Library

ISBN13: 978-1-032-15371-1 (hbk)
ISBN13: 978-1-032-15372-8 (pbk)
ISBN13: 978-1-003-24382-3 (ebk)

DOI: 10.4324/9781003243823

Typeset in Myriad Pro
by codeMantra

Publisher's Note
The publisher accepts responsibility for any inconsistencies that may have arisen during the conversion of this book from journal articles to book chapters, namely the inclusion of journal terminology.

Contents

Citation Information

The chapters in this book were originally published in the *Journal of Travel & Tourism Marketing*, volume 37, issue 5 (2020). When citing this material, please use the original page numbering for each article, as follows:

Introduction
Editorial
Mimi Li, Xiaoting Huang and Han Shen
Journal of Travel & Tourism Marketing, volume 37, issue 5 (2020) pp. 533–534

Chapter 1
The reflexive self-project of "lifestyle entrepreneurial migrants"
Jingru Zhang, Hazel Tucker and Julia N. Albrecht
Journal of Travel & Tourism Marketing, volume 37, issue 5 (2020) pp. 535–546

Chapter 2
Role Shifting Between Entrepreneur and Tourist: A Case Study on Dali and Lijiang, China
Xiaoxia Sun and Honggang Xu
Journal of Travel & Tourism Marketing, volume 37, issue 5 (2020) pp. 547–561

Chapter 3
Dining at luxury restaurants when traveling abroad: incorporating destination attitude into a luxury consumption value model
Norman Peng, Annie Chen and Kuang-Peng Hung
Journal of Travel & Tourism Marketing, volume 37, issue 5 (2020) pp. 562–576

Chapter 4
China watching: luxury consumption and its implications
Kam Hung, Hanqin Qiu Zhang, Basak Denizci Guillet and Liang Wang
Journal of Travel & Tourism Marketing, volume 37, issue 5 (2020) pp. 577–592

Chapter 5
"It is a way of life": detecting Chinese students' wellbeing during the Spring Festival homecoming
Ying Zhao, Fangfang Chen and Honggang Xu
Journal of Travel & Tourism Marketing, volume 37, issue 5 (2020) pp. 593–607

Chapter 6
Modeling the perception of walking environmental quality in a traffic-free tourist destination
Linchuan Yang, Xu Wang, Guibo Sun and Yuan Li
Journal of Travel & Tourism Marketing, volume 37, issue 5 (2020) pp. 608–623

Chapter 7

"I deserve a break!": how temporal landmarks and the perception of deservingness influence consumers' travel motivation and intention
Qing Li and Larry Yu
Journal of Travel & Tourism Marketing, volume 37, issue 5 (2020) pp. 624–635

Chapter 8

Towards the recovery mechanisms of leisure travel experiences: does the length of vacation matter?
Chun-Chu Chen, Suiwen (Sharon) Zou and Jie Gao
Journal of Travel & Tourism Marketing, volume 37, issue 5 (2020) pp. 636–648

Chapter 9

Listening to the murmur of water: essential satisfaction and dissatisfaction attributes of thermal and mineral spas
Helena Rodrigues, Ana Brochado and Michael Troilo
Journal of Travel & Tourism Marketing, volume 37, issue 5 (2020) pp. 649–661

Notes on Contributors

Julia N. Albrecht, Department of Tourism, University of Otago, Dunedin, New Zealand.

Ana Brochado, Instituto Universitário de Lisboa (ISCTEIUL), Business Research Unit (BRU-IUL), DINÂMIA'CET – IUL, Centre for Socioeconomic and Territorial Studies, Lisboa, Portugal.

Annie Chen, Roehampton Business School, University of Roehampton, London, UK.

Chun-Chu Chen, School of Hospitality Business Management, Washington State University, Vancouver, USA.

Fangfang Chen, School of Tourism Management, Sun Yat-sen University, Guangzhou, Guangdong Province, China.

Jie Gao, Department of Marketing, Montclair State University, USA.

Basak Denizci Guillet, School of Hotel and Tourism Management, Hong Kong Polytechnic University, Tsim Sha Tsui East, Kowloon, Hong Kong.

Xiaoting Huang, Department of Tourism Management, Shandong University, Jinan, China.

Kam Hung, School of Hotel and Tourism Management, Hong Kong Polytechnic University, Tsim Sha Tsui East, Kowloon, Hong Kong.

Kuang-Peng Hung, Department of Business Administration, Ming Chuan University, Taipei, Taiwan.

Mimi Li, School of Hotel and Tourism Management, The Hong Kong Polytechnic University, China.

Qing Li, Department of Hospitality Management, Shenzhen Tourism College of Jinan University, The OTC, Nanshan District, Shenzhen City, Guangdong Province, China.

Yuan Li, School of Architecture and Civil Engineering, Xiamen University, China.

Norman Peng, Department of Fashion, Marketing, Tourism and Events, Glasgow Caledonian University, UK.

Helena Rodrigues, Instituto Universitário de Lisboa (ISCTE-IUL), Business Research Unit (BRU-IUL), Lisboa, Portugal.

Han Shen, Tourism Department, Fudan University, Shanghai, China.

Guibo Sun, Department of Urban Planning and Design, Faculty of Architecture, The University of Hong Kong, China.

Xiaoxia Sun, Shenzhen Tourism College of Jinan University, China.

Michael Troilo, The University of Tulsa, USA.

Hazel Tucker, Department of Tourism, University of Otago, Dunedin, New Zealand.

Liang Wang, School of Management, Zhejiang University, Xihu District, Hangzhou City, Zhejiang Province, China.

Xu Wang, Department of Natural Resources of Sichuan Province, China.

Honggang Xu, School of Tourism Management, Xinjiang University, Urumchi, China; School of Tourism Management, Sun Yat-sen University, Guangzhou, China.

Linchuan Yang, Department of Real Estate and Construction, Faculty of Architecture, The University of Hong Kong, China.

Larry Yu, School of Business, The George Washington University, USA.

Hanqin Qiu Zhang, School of Hotel and Tourism Management, Hong Kong Polytechnic University, Tsim Sha Tsui East, Kowloon, Hong Kong.

Jingru Zhang, Department of Tourism, University of Otago, Dunedin, New Zealand.

Ying Zhao, School of Tourism Management, Sun Yat-sen University, Guangzhou, Guangdong Province, China.

Suiwen (Sharon) Zou, Department of Recreation, Park and Tourism, San Francisco State University, USA.

Introduction

Mimi Li, Xiaoting Huang and Han Shen

Lifestyle, referring to individuals' patterns of living, includes numerous dimensions such as people's interests, opinions, and activities (Vyncke, 2002). Over time, lifestyle has received increasing attention in the marketing domain for its functions in the prediction, explanation, and reflection of consumer behavior. Lifestyle is also a key factor in market segmentation and strategy development; it can provide rich information about customers compared to sociodemographic variables (Iversen et al., 2016).

Within the tourism context, the role of lifestyle is more diversified. On one hand, lifestyle influences travelers' destination-related attitudes, travel motivations, destination selection, tourism activities, and product/service consumption. Travel choices also mirror lifestyle characteristics that can shape tourists' behavior. On the other hand, lifestyle can be considered a "tourism product," as the typical mode of living (e.g., a slow-paced life) in different destinations may become an intangible attraction that draws visitors. While the vastness of lifestyle as a construct has inspired unique ideas for tourism research, related studies have not yet entered the tourism mainstream; many avenues remain underexplored. Accordingly, this special issue seeks to cast light on lifestyle's place in tourism within diverse social and cultural contexts. The chosen research endeavors clarify a variety of intriguing tourism phenomena and provide fresh insight for scholars and practitioners.

This special issue presents nine empirical papers authored by 27 researchers from Mainland China, Hong Kong, Taiwan, New Zealand, the US, the UK, and Portugal. The articles cover topics such as lifestyle entrepreneurship, lifestyle mobility, luxury experiences, and tourism well-being. This issue includes a balanced methodological set of qualitative and quantitative analyses. Unlike most research focusing on Western contexts, several of these studies involve Asian regions (particularly China, including Hong Kong and Taiwan) and thus capture the growing popularity of Asian perspectives. Selected paper highlights are summarized below.

Using ethnography approach, Zhang, Tucker, and Albrecht (2018) examined lifestyle entrepreneurial migrants' preferred living patterns. The authors discovered that these migrants' lifestyle is characterized by anti-commercialization and anti-modernity, which is a reflexive project of the self. Similarly, Sun and Xu (2019) investigated how lifestyle tourism entrepreneurs shift during daily life between enterprisers and tourists at a micro-level through in-depth interviews; results revealed several challenges that these entrepreneurs face in tourism destinations. Both studies were performed within Chinese ancient towns and focused on elements of lifestyle entrepreneurs' actual lives. Although a large body of knowledge has considered the antecedents of lifestyle entrepreneurship, the content and consequences of this topic have begun to attract new attention.

The next two papers refer to luxury consumption, first broadly and then in a more specific setting. Through focus group discussions, Hung, Qiu, Guillet, and Wang (2018) explored Chinese residents' motivations for luxury consumption along with their perceptions of Hong Kong as a luxury shopping destination. The authors concluded that Chinese luxury consumption is personally and socially oriented, and social movements may influence the appeal of a luxury destination. Peng, Chen, and Hung (2019) used a questionnaire to investigate tourist consumption in Taiwan luxury restaurants. They found that perceived value influenced patrons' satisfaction and in turn informed visitors' attitudes toward the destination and their behavioral intentions around luxury restaurants.

Lifestyle mobility is also examined in this thematic collection. Grounded in the notion of "therapeutic mobilities," Zhao, Chen, and Xu (2019) investigated the relationship between the annual Spring Festival homecoming and well-being among Chinese college students. After interviews, the authors determined that students may experience mental and physical contradictions rather than solely positive well-being upon returning home. Yang, Wang, Sun, and Li (2019) conducted a questionnaire survey to understand travelers' perceived quality of a walking system in a pedestrian-only destination. According to a proposed ordered probit model, they noted that destination crowding and cleanliness could each be improved.

Moreover, Yu and Li (2018) examined the effects of temporal landmarks on individuals' travel motivations and intentions via three experimental studies. Findings indicated that public and personal temporal landmarks facilitated individuals' motivations while internal goal conflicts could weaken the effects of these landmarks. This study sheds light on tourists' travel motivations and decision making.

The last two papers explore the association between well-being and tourism. Drawing upon online survey data, Chen, Zou, and Gao (2018) investigated how leisure travel can help people recover from work-related stress. Autonomy was identified as a fundamental link between individuals' life satisfaction and travel experiences, with vacation length exerting an additional impact on life satisfaction. In a similar vein, based on sentiment analysis of online reviews posted on Booking.com in 2015–2019, Rodrigues, Brochado, and Troilo (2019) examined factors behind consumers' satisfaction and dissatisfaction with thermal and mineral spas. Results pointed to five themes and generated constructive managerial implications related to health and wellness tourism products.

Together, the articles in this special issue present multiple perspectives on the concept of lifestyle in tourism, ranging from what different lifestyles look like to how lifestyle influences visitors. The guest editors hope this collection will pique researchers' and practitioners' interest in the lifestyle theme to encourage work featuring theoretical and methodological innovation. We also hope to see interdisciplinary academic efforts around lifestyle's role in tourism. In closing, the guest editors would like to thank all the authors and reviewers who contributed to this special issue.

References

Iversen, N. M., Hem, L. E., & Mehmetoglu, M. (2016). Lifestyle segmentation of tourists seeking nature-based experiences: The role of cultural values and travel motives. *Journal of Travel & Tourism Marketing, 33*(sup1), 38–66. https://doi.org/10.1080/10548408.2014.998359

Vyncke, P. (2002). Lifestyle segmentation: From attitudes, interests and opinions, to values, aesthetic styles, life visions and media preferences. *European Journal of Communication, 17*(4), 445–463. https://doi.org/10.1177/02673231020170040301

The reflexive self-project of "lifestyle entrepreneurial migrants"

Jingru Zhang, Hazel Tucker and Julia N. Albrecht

ABSTRACT

Lifestyle as the motivation for tourism-related migration is a prominent topic of research, but the lifestyle actually chosen, or lived, by such migrants is not fully examined. Based on ethnographic research on lifestyle entrepreneurial migrants in Dali, China, this paper considers their lifestyle as a reflexive project of the self. It identifies typical elements of such lifestyle, which is a complex combination of production and consumption, and highlights that, with the threat of increasing commercialization, some migrants reflexively reconstruct their lifestyle as a marketable business label. Their lifestyle in turn feeds into a touristic desire for "alternative" ways of life.

Introduction

"Lifestyle migration" is a relatively new concept and tends to be associated with affluent societies (Benson & O'Reilly, 2009). The concept of "lifestyle", which is derived from a sociological tradition, can be simply defined as "a distinctive, hence recognizable, mode of living" (Sobel, 1981, p. 28). In this definition, Sobel emphasizes a fundamental characteristic of lifestyle – expressiveness, which implies the existence of alternatives. If the alternatives are unavailable or constrained, the expression of a certain lifestyle is meaningless. Giddens (1991), in a similar vein, argues that lifestyle is less of an issue in traditional societies, where an individuals' life is heavily determined by their position in society. The question of lifestyle is only possible when one has the choice within a range of feasible options, which is the very situation of many, though not all, people in late modernity.

Previous studies on the topic of lifestyle migration are predominantly Western-oriented (Benson & Osbaldiston, 2014). However, benefiting from economic reform, China has not only experienced an economic boom in recent years, but has also witnessed a "mobility shift" (Xu & Wu, 2016). This shift is from production-led peasant worker migration to lifestyle migration (Xu & Wu, 2016). In contemporary China, such migration increasingly involves moving to a tourism destination and opening a small tourism business. These urban-to-rural migrants can therefore be considered "lifestyle entrepreneurial migrants" (LEMs), and the main motivation for this type of migration appears to be a search for a better way of life rather than making financial profits (Liang, Xu, & Thomas, 2010). Such a move is regarded by many people, and particularly those living in urban areas, as an ideal "lifestyle" choice.

According to Levy, lifestyle is "a large complex symbol in motion" (1999, p. 218). To choose a certain lifestyle is thus to express the personal preference or attachment to the symbolic meaning embodied by it. In this sense, the choice is not only concerned with "what to do" but also "who to be", and self-identity becomes the basis of the choice-making. Furthermore, in Giddens's (1991) theorization of self-identity, reflexivity plays a fundamental role in the formulation of self in late modernity. Reflexivity is the ability to consider oneself in relation to his/her (social) context and this may occur most poignantly when there is a felt absence of social guidelines for actions in novel situations (Archer, 2007). In modern societies, dynamism and pluralism undercut the authority and certainty which were present in pre-modern society, so that in modernity, and even more so in "late modernity" (Giddens, 1991), individuals are compelled to engage in the ongoing task of reflexively redefining their situation and reorganizing their own biography. With the decline of authoritative guidelines, individuals' biographies are less prescribed than before and a vast range of optional self-identity components become available (Archer, 2007; Giddens, 1991). Self in late modernity is therefore constructed largely by one's own self-determining choices, and lifestyle, which filters, integrates and routinizes these choices, can be seen as the material form of a particular narrative of self-identity. In other words, to build a lifestyle means to "work on" one's self, and so in this sense, lifestyle can be regarded as a product of "reflexive project of the self" (Giddens, 1991).

As a key concept of entrepreneurial migration to tourism contexts, "lifestyle" has thus far predominantly been studied in relation to the motivations for migration (Sun & Xu, 2017; Wang, Hung, & Bao, 2014). Less attention has been paid to the kinds of "lifestyle" which are actually chosen and "lived" by these migrants nor, in other words, to the idea of lifestyle choice as a "reflexive project of the self". This article addresses these chosen and "lived" aspects of lifestyle migration, and does so through an ethnographic analysis of LEMs' everyday practice in Dali, China. The research aims to identify typical elements of this chosen lifestyle and to thereby highlight its *expressiveness* (Sobel, 1981). Doing so enables an understanding of LEMs which goes beyond consideration only of the lifestyle "sought", and adds the notion of "self" re-formation, or "self-project", via the particular chosen and lived lifestyle. As a relatively new opportunity in contemporary China, and one afforded specifically in certain peripheral tourism contexts, research into lifestyle tourism-related entrepreneurial migration in China is important in order to understand the role that such migration plays in China's manifestations of "late modernity", and also with regards to formations of certain individuals' identity-making. With this understanding as the main purpose of this paper, then, the paper begins by outlining extant literature related to lifestyle migration in tourism contexts, and then it will go on to discuss the ethnographic research and its findings in relation to the ways in which the "lifestyle" of LEMs in Dali constitutes a reflexive self-project.

Lifestyle migrants in tourism contexts

In recent decades, the notion of lifestyle migration has attracted the interest of researchers from increasing variety of disciplines, such as geography, tourism, and sociology (Benson & O'Reilly, 2009; Cohen, Duncan, & Thulemark, 2015; Xu & Wu, 2016). To have a better understanding of this phenomenon, researchers suggest to integrate it into a broader perspective of mobility or migration. In Benson and O'Reilly's definition (2009, p. 609), lifestyle migrants are "relatively affluent individuals of all ages, moving either part-time or full-time to places that, for various reasons, signify, for the migrant, a better quality of life". They suggest that if one relocates to change one's lifestyle, one can be called a lifestyle migrant. In their theoretical framework, lifestyle migration should be considered as embedded within two broader circumstances of movement. The first is tourism, which constructs and markets a "better" lifestyle for lifestyle migrants and offers them a feasible and attainable way to achieve it. The second is counter-urbanization, which breeds sentiments of anti-modern and anti-urban, constructing the "rural idyll" as a better way of life.

Lifestyle migration is usually regarded as a sub-category of tourism-related migration. Tourism-related migrants, Yang and Lu (2008) argued, are individuals who relocate to tourism destinations temporarily or permanently to work in the tourism industry, or to enjoy the same environment as tourists. The typology of tourism-related migration is usually based on a binary structure: production-led migrants and consumption-led migrants (Williams & Hall, 2002; Yang & Lu, 2008). In this typology, the former refers to individuals who are motivated to migrate with the intention of engaging in production activities in the new destination, and can be further subdivided into entrepreneurial migrants and labor migrants. On the other hand, consumption-led migrants refer to individuals who primarily seek leisure or health in their new place of residence, and can be generally considered as lifestyle migrants (Williams & Hall, 2002). A further typology developed by Tang and Xu (2015) supplements a new type which represents the intersection of this duality: lifestyle entrepreneurial migrants, the ones who are mainly driven by consumption motivations (to seek a better way of life) but also are involved in production activities in the destination.

The phenomenon of lifestyle entrepreneurial migration was initially observed by Williams, Shaw, and Greenwood (1989), who argued that in order to maintain the new, desirable lifestyle in destinations, many migrants, especially young people, have to engage in production activities to generate income. To achieve a better balance of life and work and gain more freedom, LEMs prefer to run their own small businesses, often related to the tourism industry (Benson & O'Reilly, 2009; Stone & Stubbs, 2007). This is partly because there are relatively low market barriers to running small tourism businesses, and it requires less general and professional skills and knowledge than other industries (Ateljevic & Page, 2009).

Hence, when discussing entrepreneurial motivations, studies have tended to focus on the differences in working conditions between origin and destination places. Running a lifestyle business is seen as allowing migrants to gain a sense of personal freedom, independence, a flexible schedule, and a better work-life balance. Furthermore, commercial and lifestyle oriented motivations are a significant dichotomy for lifestyle entrepreneurs (Bredvold & Skålén, 2016; Shaw & Williams, 2004). Dewhurst and Horobin (1998) developed a continuum between these two motivations, and indicated that they are not mutually exclusive. Most lifestyle entrepreneurs do not seek growth and profit maximization but use the income to maintain their lifestyle. However, previous studies indicate that their initial motivations and goals can change to be more profit-oriented due to the commercialization of destinations (Wang et al., 2014).

Overall, then, while "lifestyle" is the foundational concept of LEM, it has tended to be treated as a premise, or rather as a motivation, in the related research. Closely linked to the motivation for business, LEM's motivation for migration is also another frequently discussed issue. Pull and push theory is often used to identify LEM's motivations for migration (Chen & Yang, 2014; Sun & Xu, 2017; Walker & Brown, 2004). Push factors usually refer to negative reasons from the origin, including pollution, a feeling of alienation, and high living costs, while pull factors refer to positive "pulls" towards the destination, including climate and health benefits, beautiful scenery, community spirit, and low living costs. Unlike Western LEMs, Chinese LEMs usually look for their destinations at home rather than overseas (Sun & Xu, 2017).

Although lifestyle as the motivation for LEMs is an important topic of research in this field, the constitution of LEMs' achieved, or lived, lifestyle is not so fully examined. Moreover, whilst the concept of "lifestyle entrepreneurial migration" is a combination of consumption and production, and so it follows that "work" should be included as an important element in these people's lifestyle choices. However, the word "lifestyle" in contemporary society generally refers to "consumer lifestyle" (Binkley, 2007), and production activities are frequently excluded from the discourse of lifestyle. Since production activities are considered to be less subject to individual agency and more dominated by economic compulsion (Giddens, 1991, p. 81), it is generally understood that in the production realm there is a lack of "choice" on how to express oneself. In the context of lifestyle entrepreneurial migration, however, a self-determining "choice" comes very much to the fore. For example, researchers have already found that LEMs actively engage in alternative production activities different from traditional business modes (Bredvold & Skålén, 2016; Xu & Wu, 2016).

Taken the notion of self-identity in late modernity as constructed by one's self-determining choices, lifestyle as a choice can be viewed as the material manifestation of a particular, chosen, narrative of self-identity. In other words, as lifestyle inherently implies alternatives, LEMs could be regarded as an ultimate way to pursue Gidden's notion of a "reflexive project of the self". Yet, due to the predominant focus on lifestyle as motivation in the extant research, there is a lack of understanding regarding the role of lifestyle entrepreneurial migration vis-à-vis matters of self-identity, and more particularly, the ability to "choose" and re-form one's self-identity, in contemporary societies. Indeed, Giddens's (1991) argument is that in contemporary society, people do not adopt a lifestyle unreflexively. On the contrary, they choose elements which can serve their self-identities and design them together into a "lifestyle" (Featherstone, 1987). Hence, this research set out to ascertain what kind of lifestyle is actually "chosen", achieved and "lived" by LEMs in Dali in order to gain a fuller understanding of contemporary "projects of the self", particularly as they may manifest in contemporary Chinese society.

Study method

Interpretive qualitative research is the underlying methodological design of this article. Since Chinese LEMs prefer destinations at home rather than overseas, destinations, like Dali, Lijiang and Yangshuo, have become popular communities for them (Wang et al., 2014; Xu & Wu, 2016). One significant common characteristic of these places is that they are all located in the socio-economic periphery of China so that they offer a comparatively alternative atmosphere, compared to contemporary cities. For this research, Dali was selected as the field site and a six-month-long period of ethnographic fieldwork was undertaken, from May to October 2017. Dali is a historical town in southwest China, which has over 1000 years of history. It is also an autonomous prefecture of Bai ethnicity, which has developed a vernacular culture distinct from the traditional culture of Han Chinese. Besides its rich historical and cultural resources, it has a mild climate and spectacular scenery, which together make it a top destination for Chinese tourists as well as LEMs (Liang et al., 2010; Ma & Xu, 2016; Xu, Ma, & Jiang, 2017).

While against the positivist view of the relations between the research and the researched, the authors considered the interaction with subjects was more like a process of negotiation than discovery, which dismissed the "independent" or "uninvolved" position of researchers and encouraged them to embrace their own "horizons" (Gadamer, 2004). Participant observation and semi-structured interviews were employed as information-gathering methods. In order to gain insight into LEMs' everyday lives, the first author lived in a guesthouse run by two respondents during the period of the research. Accessing various circles of LEMs in Dali, the author spent most of his time with LEMs participating in their working and leisure activities. Their daily practices as well as the author's reflections were recorded as ethnographic field notes. In addition, thirty-four LEMs were formally interviewed, with an average duration of two hours. Snowball sampling combined with maximal variation sampling were applied. The interviews were held in respondents' working places (such as their hostels or cafe shops) and recorded with the consent of respondents. Among the thirty-four LEMs, there were eleven females and twenty-three males with their ages ranging from 25 to 49. Their current length of residence in Dali ranged from one month to nine years (see Table 1).

Adopting a thematic analysis approach, the recordings of interviews and field notes were analyzed with the use of Nvivio 11. Boyatzis (1998) contended that thematic analysis can be considered as a continuum from theory-driven and data-driven approaches. The current article took a middle position. On the one hand, as discussed above, the concept of "lifestyle" is considered a combination of two basic components: production and consumption, so three theoretical themes – "lifestyle" and its two sub-categories "production" and "consumption" – were preassigned. On the other hand, the empirical material was coded in a data-driven approach to developing other categories. First, as suggested by O'Reilly (2012), all the transcripts were read through to gain a basic sense of potential themes and categories. Second, open coding was conducted sentence by sentence so as to reduce the raw data into a manageable number of codes. Third, focused coding then refined the developed codes. According to the analysis, the preassigned theme "lifestyle" was modified into two themes – "lifestyle as expressive self-project" and "lifestyle as marketable self-project". Each of them consists of two categories – production and consumption. The remaining subcategories were assigned into these categories respectively. Figure 1 illustrates the themes and (sub)categories yielded by the analysis.

After developing the thematic codes, the study took "thick description" (Geertz, 1973) as the main principle to interpret the themes. This requires not only to identify the behavior patterns, but also to explain them in their social and cultural contexts. The meaning-making through these behavioral patterns, linked with the reflexive project of self, were explored to understand the lifestyle chosen and expressed by LEMs, and also how the chosen lifestyle becomes a marketable lifestyle in the tourism context. Taking an involved position of the interpreters, the validity of interpretation is against the positivist view of truth. This positioning implies that the truth of a claim is based on the consensus reached by a cultural community rather than objectivity (Carspecken, 1996). Thus, the validity of this research is also based on the negotiation and consensus between the researchers and the researched. Besides long-term fieldwork and triangulation, member checking with LEMs was adopted to negotiate a consensus of meaning. The key findings will now be outlined in the next sections.

Findings

Lifestyle as expressive self-project – expressive "production self"

Most of the LEMs in Dali would agree that the state of work in Dali is more ideal than in previous places of residence. As Marx (1998, p. 37) put it, "[the mode of production] is a definite form of activity of these individuals, a definite form of expressing their life, a definite mode of life on their part. " As the involvement in tourism businesses, such as running a hostel or cafe shop, is a

Table 1. Profile of respondents – lifestyle entrepreneurial migrants.

Name	Gender	Age	Date of Migration	Occupation	Name	Gender	Age	Date of Migration	Occupation
Rui	Female	28	2017	G	Luo	Male	28	2017	G
Fei	Female	35	2012	G	Xiang	Male	49	2012	G
Lang	Male	23	2016	G	Qi	Female	27	2014	G
Guan	Male	32	2014	G	Ping	Female	40	2013	G & R
Ma	Male	39	2013	G	Tong	Male	28	2012	G & R
Chai	Male	44	2009	G	Sun	Female	35	2015	G & R
Ming	Male	35	2013	G	Bei	Male	28	2013	R
Lao	Male	35	2010	G	Huan	Female	28	2014	R
Ji	Male	29	2014	G	Dun	Male	45	2014	R
Yu	Male	37	2013	G	Yang	Male	30	2015	R
Song	Male	47	2017	G	You	Female	43	2012	R
Cen	Female	39	2016	G	Miao	Female	27	2014	O
Peng	Male	30	2017	G	Jiao	Female	25	2016	O
Shang	Male	36	2016	G	Zhuang	Male	32	2017	O
Cang	Female	35	2016	G	Cha	Male	45	2012	O
Chuan	Male	34	2011	G	Yue	Male	30	2014	O
Dan	Male	28	2016	G	Bai	Male	22	2015	O

* Occupation: G = owner of guesthouse; R = owner of restaurant/café shop; O = owner of other tourism business

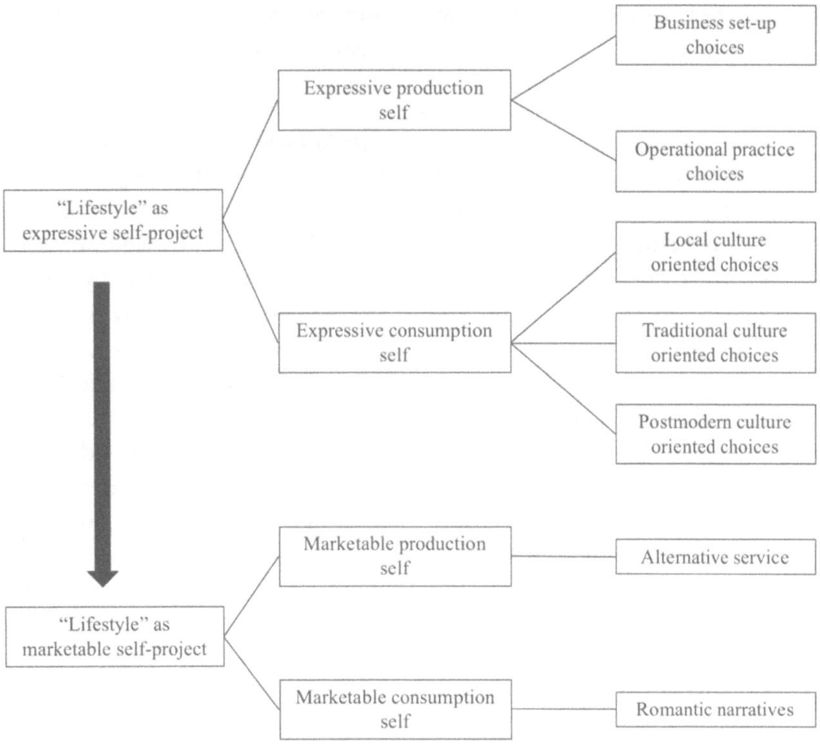

Figure 1. Thematic analysis of LEMs' lifestyle.
LEMs: lifestyle entrepreneurial migrants.

voluntary choice for these people and the mode of self-employment guarantees the high degree of control over the work, their production practice in Dali can be considered as a significant part of effort to reconstruct their life trajectories (Giddens, 1991). In addition, heavily shaped by a sentiment of anti-commercialization, LEM's production practice is also entwined with the production of, and expression of a particular "alternative" self-identity. As a respondent commented on his business, "This is what I believe my individuality is. This is what I believe my guesthouse should be" (Tong, male, 28). As such, LEMs' production practice can be divided into various "choices" which then work towards the production and expression of an "alternative" self. These choices can then be grouped, albeit loosely, into "business set-up" choices and "operational practice" choices. These are laid out below.

"Business set-up" choices

Location selection. The first issue or choice relating to business set-up is that of location selection. According to general business practice, various factors need to be taken into consideration when selecting a location, among which traffic convenience and accessibility is one of the most frequently mentioned principles (Yang, Wong, & Wang, 2012). Conversely, however, lifestyle enterprises are usually located in peripheral tourism areas in Dali, such as non-trunk roads in the town or villages nearby the town. Two of the respondents participating in this study had even located their guesthouses in villages which were more than 10 miles away from the town and where there was no other tourism business and facility around. This was explained by one entrepreneur as follows:

> To earn money is not our main purpose. The most important thing is to do what you like and live a happy life. So, to be frank, the site selection for our guesthouse is not proper [for business]. The site is on the outside of Dali Old Town, and even on the edge of this village. However, its yard is very big and quiet. We are very happy to live here. That's the reason we chose this site. (Fei, female, 35)

For LEMs, the work place is usually also their place of living. The spatial combination of these two functions leads to particular decisions regarding location selection. Considering "to live a better life" as the first principle, they tend to emphasize having a spacious house for themselves (not for tourists), a quiet neighbourhood (not the area with a large flow of tourists), accessibility to living facilities (not to tourism facilities), and low cost (rather than locating for higher revenue). Therefore, the site selection of their business is expressive of "I-like-it" rather than "customers-like-it". In other words, the location choice for their business is based more on their own "lifestyle" (self-focused), rather than selecting what would work best for tourists and business per se (customer-focused).

Do-it-yourself. Another notable business set-up choice is that many LEMs prefer the guesthouse or restaurant building to be in a run-down condition. This is not only for cost reasons, but also it provides them with the opportunity for "do-it-yourself" restoration or decoration work; "If you live in Dali, you will be finally transformed from a green hand into a locksmith, plumber, electrician..."(Rui, female, 28). Moreover, LEMs appear to want to undertake the design, the decoration, and to carry out repairs by themselves rather than engage workers (Figure 2(a,b)). This desire for DIY among LEMs is partly due to the lack of this kind of experience in their previous lives. Since the Chinese economic reform in 1979, marketization has penetrated even the most domestic aspects of Chinese people's daily life. DIY has become an uneconomic and inefficient thing in contemporary society due to the low labor costs in the Chinese market. In this sense, then, DIY represents to LEMs a more "traditional" lifestyle in a commercialized society and it provides people with the opportunity to become what they consider a more "traditional" person. Furthermore, LEMs' narratives emphasize the "fun" and "leisure" aspect rather than the "work" aspect of DIY practices: "The most attractive thing of this house is that it needed to be renovated. If everything is ready, it loses the fun of do-it-yourself" (Ku, female, 30). This echoes some researchers' finding that DIY practice in postmodern society have shifted its emphasis from the utility of its end products to its pleasure, expressiveness, and communicative function (Tanenbaum, Williams, Desjardins, & Tanenbaum, 2013). The notion of "fun" in DIY implies the "non-productive" feature of LEMs' life and, in some way, constructs the identity of a "new leisure class" (Veblen, 1924) in China.

Operational practice choices

Individualized theming. Firstly, the operational practice of LEMs' businesses is characterized by individualized theming. Lifestyle-oriented businesses are found to be strongly branded by the individual character of the owners (Bredvold & Skålén, 2016). As Ma and Xu (2016) have also previously noted, rather than run what they consider to be a bland, featureless business, LEMs prefer to combine their businesses with personal interests, such as fitness, handicraft, cycling, or literature, and this helps the development of individualized themes. For example, a couple among the respondents run a guesthouse themed on the "parent-child relationship". This idea stemmed from their life experience before migrating to Dali. As the parents were too busy to accompany their child when they living in Beijing, they experienced a form of "family in crisis". To help resolve their sense of the Chinese family being in crisis, they decided to migrate from Beijing to Dali and run a guesthouse which could provide them with more free time to stay with each other. To highlight the "parent-child" theme of the guesthouse, they provided 4 family rooms out of 11 and increased public room and facilities for children. Moreover, they always invited their guests to have tea and conversation with them in order to share the experiences of children's education and upbringing. They considered this as the unique culture of their guesthouse, a genuine expression of their life experience, and they opposed this to the so-called corporate culture of hotels which they believed to be "inauthentic" and "insincere". As the wife put it, "I think the biggest difference between hotels and our guesthouse is that hotels don't have a 'culture'. You just pay

Figure 2. (a) The DIY restoration of an old cottage (before, taken by Rui). (b) The DIY restoration of an old cottage (after, taken by the first author).
DIY: do-it-yourself.

money and stay there. You'll never have a cup of tea with the owner."

Interaction with customers. The above point about owners' interaction with their customers is another notable practice for LEMs' business operations. It is also a further way in which the work-leisure mix features in LEMs' daily practice. Many LEMs themselves are keen to interact with a wide variety of people. They are keen to invite their guests to enter the "backstage" (MacCannell, 1973) of their daily life, for instance, to have a chat over tea or beer, or to join a dinner or a party with the owners and their friends: "Some guests became my friends. After they went back to the city, we would chat on the Internet. Sometimes they even mailed local specialties to me" (Chai, male, 44). As new guests continuously visit their guesthouses or cafe shops, their businesses become an ideal platform to connect to the outside world and make friends, as noted in previous research (Ma & Xu, 2016).

What also emerged from the present research is how, in a sense, the guests are also the audiences of LEMs' everyday performances. Through the interaction, LEMs express and reinforce their new self-identity, which is embodied in an elaborately designed lifestyle. This also feeds into the guests' desire for an "alternative" life. As one LEM put it, "You are a model who live a free life in their imagination" (Ma, male, 39). Moreover, LEMs recognize that interaction with customers also has its productive function. In the context of lifestyle-oriented businesses, by showing the backstage of life, the owners provide guests with an environment with interpersonal authenticity, or in other words, provide an intimate "host-guest" relationship. This conceals the essence of the commodity exchange between them, and thereby widens the customers' zone of tolerance by lowering their acceptable level of service. "When you become friends with the guests, they won't mind if you make a mistake" (Ji, male, 29).

Anti-professional operation. Another notable characteristic of LEMs is the non, or anti-professional character of their business operation. Most LEMs have little previous knowledge of how to run a tourism business, and they frequently overlook, or even reject, the professional industry knowledge and marketing strategies which could serve their business better. For example, as online travel agencies (OTAs) play an increasingly important role in the distribution of tourism products and services (Park, Gretzel, & Sirakaya-Turk, 2007), small tourism businesses are heavily dependent on them. Many owners recognize the impact of online reviews on consumers consideration, and manipulating online reviews becomes an unspoken rule among owners. However, many LEMs express a strong dislike for this behaviour: "I think it's fake, so I don't want to do that… I know I should promote my guesthouse through social media, but I don't. My opinion is that it's all up to you whether you come or not" (Sun, female, 35). Even though LEMs know that the deliberate rejection of this common business practice will lead their business towards an unfavorable position when competing with other tourism businesses, they persist with these "anti-professional" operations. This can be seen as an expression of their sentiment of anti-commercialization, and so for them, the maintenance of this identity seems more important than the seek for material wealth. As one LEM expressed:

> I don't just consider it as a business, because it is also a place where I live and play. I never expect many guests to come…At the worst, even if this guesthouse is forced to close, I can still live here as my home (Cen, female, 39).

Lifestyle as expressive self-project – expressive "consumption self"

As Baudrillard (2001, p. 22) put it, "Consumption, in so far as it is meaningful, is a systematic act of the manipulation of signs." Individuals express and construct their identity through consuming these "signs". In this view, consumption should be understood from a symbolic rather than material perspective (Levy, 1999). The "consumption self" of LEMs in Dali expresses a sentiment of anti-modernity, which is particularly fed by three streams of cultures: "local" culture, "traditional" culture and "postmodern" culture.

Local culture oriented choices

Many LEMs in Dali endeavor to "look local", as represented by local Bai ethnicity. Like tourists, LEMs often wear ethnic-styled costume or ethnic ornaments such as silver bracelet with ethnic patterns. However, this so-called ethnic-styled looking is different from the "authentic" dress style of the Bai people. As one LEM noted:

> [This style] integrates the merits of costume from various ethnicities and becomes what it is. It's just mix-blood. The ethnic-styled costume in Dali has its distinctive style…But one thing you may notice is that only those migrants would dress like that. Local people rarely dress it. (Zhuang, male, 32)

Dress, as one of the bodily practices, is an important symbolic boundary between self and other and a visible commitment to the membership of a sub-culture, therefore it helps to perform and reproduce identity (O'Regan, 2016). In this case, the ethnic-style is different from either urban style or Bai style, which suggests that LEMs endeavour to differentiate themselves from people in the modern city, that is the people they used to be, and from local people at the same time. As such, they form a distinctive group who are easily identifiable through their style of dressing, and this dress style is a symbol of the "new Dalinese".

Traditional culture oriented choices

"Traditional culture" is another important value orientation for LEMs and specifically refers here to Chinese classic traditional culture. As Dali is a historic town with over 1000 years of history, the traditional physical environment and sense of community provoke a strong sense of nostalgia. Compared with their previous lives in cities, LEMs tend to think their lives in Dali are more rustic and traditional. As one LEM put it:

> I think we all have a dream to live a pastoral life, where we can have our own small patch of land and achieve self-sufficiency. This perhaps is the perfect state, the highest state of life – a pastoral life like Tao Yuanming lived. (Tong, male, 28)

Tao Yuanming (365–427) is a cultural icon in China. He is well-known for his practice of reclusion from the office and his poetry about the reclusive life in the countryside. His biography is regarded as the model of Chinese recluses and his work "Peach Blossom Spring" becomes the utopia of the agrarian community in every Chinese heart (Kwong, 1989). "Here are poems to recite, wine to drink, a field to plough, a garden to dwell, a cottage to live. Here is where you give and take." This is an advertisement of a guesthouse posted on the social media and can be seen as an echo of the poetry and the life of Tao. In some sense, LEMs can be seen as the modern variant of the traditional Chinese recluses. They resign from what are generally considered to be good jobs in the cities and choose to come to Dali and live a free and simple life in a traditional community. To strengthen this feeling, they may grow their own vegetables or develop some classic traditional interests to construct a more systematic traditional life. Tea ceremony, calligraphy, and classic Chinese music are popular among LEMs.

Postmodern culture oriented choices

What can be considered "postmodern culture" is mainly featured as a "hippie culture", which is adopted as a strong influence on LEMs in Dali. In the early stages of the Chinese economic reform, the Chinese government encouraged inbound tourism rather than domestic tourism. Dali was among the first batch of destinations open to foreign tourists in 1984, and became the centre of backpacker culture and also a gathering place of foreign hippies in the old town at that time (Notar, 2006). Although it is now difficult to track it down that why these hippies gathered in Dali rather than other destinations, there is at least one significant reason – the easy accessibility of marijuana in Dali. Because of the appropriate climate in Dali, one can easily find wild marijuana in the fields. "It's Free…You can make it by yourself. There is a lot of wild weed in the mountain. Most foreigners take weed. As it's wild, it's not strong" (Cai, male, 27). Attracted by these foreign hippies and their apparently "free" way of life, Dali then became a utopia of their Chinese counterparts. Many of the respondents in this research admitted that they took marijuana in Dali: "My friends came to my guesthouse and started to smoke weed. They asked me to take one puff. I feared that I couldn't blend into them, so I said OK. Then I just took a small puff…" (Ye, male, 33). Besides the drug culture, the influence of hippie culture also can be found in their other daily appearance, such as their clothes and hairstyle, such as dreadlocks: "Dali is a fantastic place. It has the largest number of weirdos in Yunnan. All the weirdos are gathering here!" (Long, male, 27).

Lifestyle as marketable self-project

> Our business got worse and worse, but none of us cared about it at first. Tong was playing video games all the day. Once a guest called him to ask if there was any vacancy, but he answered, 'Our guesthouse is a little far from you. You can ask the others around you and find one to stay'… At last, we only earned about 1000 RMB per month, no more than the income of the cleaner we employed. So I said we must change (Xiang, male, 49).

The excerpt above is from a LEM who recalled the commercial failure of his lifestyle-oriented guesthouse. This story is the epitome of LEMs' casualness on management. Some of them change their business hours when they want, while some refuse to serve customers they don't like. The unstructured and informal way of management, coupled with their lack of professional tourism knowledge, becomes a prominent reason for their high chance of business failure. In addition, the increasing commercialization of the destination, increasing rent and increased competition are contextual challenges for LEMs with regards to their ability to maintain their "ideal" lifestyle. All of these factors force LEMs to make changes. Some LEMs change their lifestyle businesses to traditional businesses by adopting professional management. However, for those who want to maintain a consistent "alternative" narrative of self, two main options are chosen. The first is to migrate to another less commercialized and competitive destination in order to preserve the lifestyle they desire. The second option is to stay in Dali and reconstruct their lifestyle as a marketable self-project. This involves reflexively recognizing and refining the key elements that can meet touristic desire in their lifestyle, and then transforming those into marketable resources for their business.

The shift from spontaneous to self-conscious practice implies that these LEMs reflexively, consciously and purposefully "make" lifestyle, rather than merely "follow" it (Beck & Beck-Gernsheim, 2002). This "making" of lifestyle also implies a reflexive, conscious and purposeful "making" of self (Giddens, 1991), which is then marketable as a part of their tourism business product. As the lifestyle of LEMs, together with their "self-project", consists of two components – production and consumption, this reflexive (re)construction of lifestyle and self can be examined in these two perspectives.

Marketable "production self"

Production self of LEMs is expressive of its "alternative" way of service. As discussed above, one of the most significant features of LEMs' management is its casualness. On the one hand, it is a barrier for successful operations. On the other hand, it is also a distinctive point of the cultural capital of lifestyle enterprises as compared to their traditional counterparts. This special quality attracts a niche market who are apparently bored with what may be perceived as institutionalized and standard service. This kind of "alternative" service provides tourists with unexpected and novel experiences. For example, the following is a notice posted in a lifestyle-oriented restaurant in Dali:

> If you want to have a meal in this restaurant, you need to take your order, bowls and chopsticks, drinks and courses all by yourself. In addition, after you finish the meal, you also need to clean the table, pass the dirty dishes to the kitchen, and take leftovers to the rubbish bin by yourself. Dali is where to live slowly, to serve the food slowly, to have the meal slowly, and to wait for a vacant seat slowly.

In this interesting notice, the owner invoked a "life philosophy" – to live slowly – in order to persuade customers to accept this alternative service. A consumer's review on the internet demonstrates the success of this strategy: "The so unbelievable way in other restaurants comes naturally here. This non-conformist personality becomes a legendary brand. People in Dali must know about this place. Many tourists just come to see how it is." By word of mouth, the restaurant then comes to stand out from its competitors, not because of the quality of food, but because the "lifestyle" and "self" of its owner are projected.

Marketable "consumption self"

To identify the marketable elements in LEMs' consumption self and practices is to identify the elements which differentiate from modern city life. In contemporary China, "Escape from BSG (Beijing, Shanghai, and Guangzhou, which are seen as the first-tier cities in China)" has become a popular slogan among Chinese white collars (Hu, 2013). LEMs' migration to Dali can be seen as part of this movement of counter-urbanization. Their narratives of their life in Dali feature core concepts such as "escape", "freedom" and "nature". Many LEMs are aware of a similar anti-urban and pro-rural sentiment among tourists to Dali, and then consciously depict their everyday life on social media platforms in order to market their business in a seemingly inconspicuous way. Among the respondents in this study, one keeps an online diary, two create cartoons series of their daily life, and many of them regularly post photos or prose in social media. As one respondent remarked:

> I will post our life on social media...such as cooking, traveling and playing games with our guests. They are parts of our life. Anything manifesting its beauty can be posted...I will intentionally select those things to post, of course. I hope the things I show to the public are good. But I don't post them only in a promotional purpose. I think these are what I can share with others (Chuan, male, 34).

Through social media, many LEMs develop their own followers, who are identified with their lifestyle and life views. More importantly, these followers subsequently become their potential customers. This in turn makes LEMs less dependent on online booking or travel agencies. A person-to-person business mode has been established, wherein both the chosen lifestyle and the "self" of the LEM becomes the core attraction for their customers.

Conclusions

This study set out to explore the kind of lifestyle that is chosen, and lived, by LEMs in China in order to consider the ways in which lifestyle entrepreneurial migration is a late modern reflexive project of the self. As discussed above, the "lifestyle" of LEMs should not be considered as a taken-for-granted premise but rather as a certain lifestyle actively chosen among various alternatives. To commit to a chosen lifestyle means other options are seen as "out of characters" with it (Giddens, 1991). The concept of "choice", therefore, is an important perspective to understanding LEM's life, and it links with the notion of lifestyle as a self-identity project. Generally speaking, the chosen lifestyle of LEMs in Dali is expressive of the sentiment of anti-commercialization and anti-modernity, and it is so in a number of ways.

One way in which the LEMs' chosen lifestyle expresses anti-modernity is in the blurring of a production-consumption divide, which is distinct from the lifestyle which LEMs left behind in the cities they came from, where work and leisure are experienced as more significantly distinct from each other. Indeed, the lifestyle of LEMs is characterized by the complex combination of production and consumption in two aspects: (1) The spatial-temporal combination – people spend most of their work and leisure time in the same place; (2) The functional combination – people's daily practice such as DIY and the interaction with customers is usually considered as both a work requirement and a personal interest. This combination mediates the conflict between work and leisure which exists in contemporary urban society (Thompson & Bunderson, 2001). In modernity, characterized by labor division and urbanization, people suffer fragmented experience in their everyday life (Desforges, 2000). As Lefebvre (1991, p. 40) criticized the modern society, "We work to earn our leisure, and leisure has only one meaning: to get away from work." The lifestyle chosen by LEMs, on the contrary, shows a kind of "wholeness" of everyday life, whereby they work around their living places and regulate the work-leisure rhythm independently, and then maintains the "wholeness" of self in modern society, which can offers people the "ontological security" (Giddens, 1991). In fact, the alliance of work and leisure has been noticed in other lifestyle mobilities as well (Cohen et al., 2015). This research therefore provides an alternative perspective to understand this combination by establishing the link between "lifestyle" and "self-identity".

While the "wholeness" of this chosen lifestyle conveys its anti-modernity project, its elements provide more detailed meanings to strengthen this discourse. For production practice, while previous research has found that LEMs have different business features from traditional ones (Liang et al., 2010; Xu & Wu, 2017), this study legitimizes their production practice as a component of "lifestyle" by demonstrating its "expressiveness". The mode of self-employment assures LEMs' autonomy over their work and the possibilities to express their distinctive selves to traditional entrepreneurs. Modern work ethic and institution have very limited constraints on these LEMs. The sentiment of anti-commercialization becomes the central tenet of their productive practice choices.

In terms of consumption practice, three cultural streams are identified as being a part of the LEMs' reflexive self-project. As the sense of unity is an inherent tendency for self-identity (McAdams, 1997), individuals are inclined to form an integrative life narrative and represent their lifestyle in a unifying and consistent way. Thus, self-identity is a useful clue to identify the ordered patterns of lifestyle from people's trivial practices. The three streams of LEMs' consumption practice all stand opposed to "modern culture" and become the main identifiers used to reconstruct LEMs' new self-identity in a way which distinguishes them from "modern" urban people. The modification of local culture helps them become the "new Dalinese", and a traditional reclusive culture which is revered in Chinese society contributes to the justification of their new self-identity and lifestyle. The "imported" postmodern culture provides them with another way to rebel against the mainstream values of Chinese contemporary urban society. Therefore, through these chosen elements embodying different cultural genres, LEMs develop a seemingly unified lifestyle, which is purposefully opposed to the modern urban lifestyle.

However, as LEMs' lifestyle is fundamentally dependent on the tourism industry, which itself is a product of modern society (Wang, 2000), the rebelling of these LEMs against modernity is not always thorough. With the commercialization of destinations, the presentation of LEMs' lifestyle is reflexively reconstructed into a marketable one for their businesses and feeds into a "touristic" desire to experience "alternative" ways of service and "non-modern" ways of life in places such as Dali. This touristic desire is rooted in a larger scaled discourse of romantic critique of modern society (Löwy, 1987; Wang, 2000). Commercialization, modernization and urbanization are three main critical issues in this discourse, which consequently breed a sentiment of escape from cities (Hu, 2013) and allows LEMs to target a particular niche tourist market. As such, this marketable lifestyle is not only the choice of LEMs themselves, but also an expression for tourists.

While the role of tourists in the reconstruction of LEMs' lifestyle has been relatively overlooked in previous research, the transformation of LEMs' lifestyle into a key resource for their business marketing also shows some implications for industry practices. The effective "lifestyle marketing" of LEMs implies that it is the symbolic meanings, rather than the functional characteristics, that develop attachment and commitment for consumers (Saviolo & Marazza, 2013). It asks LEMs to articulate the aspects of their lifestyle which tourists identify with and are attracted to. In other words, the symbolic meanings of LEMs' lifestyle are derived not only from the expression of their own self-identity, and but also from the expression of tourists' desire, or more generally, the broader trends of a specific social-culture ethos. Moreover, the lifestyle marketing in tourism should effectively utilize the "wholeness" of lifestyle. That is, lifestyle owners should take not only the way of consumption but also the way of production (such as the way of service) as the resource for lifestyle marketing. In addition, the research also suggests that the promotional way of lifestyle enterprises is experiencing a shift from traditional distribution channels to more individualized social media, which is a more proficient platform to represent the unique meaning-making of their lifestyle.

Finally, since it is the mode of production that makes the lifestyle possible, further research might usefully explore what it is exactly about the nature of tourism business that renders it so promising as a mode of lifestyle-oriented production. Another pertinent research question stemming from the present research is to explore how the newer, younger LEMs differ in their value orientations relative to the older, more long-term LEMs. Moving forward, the growing phenomenon of "lifestyle entrepreneurial migration", as considered here as a contemporary reflexive project of the self, especially when entwined with tourism contexts, has been shown here to be an area of considerable interest in terms of its dynamic place and role in contemporary Chinese society, and as such is ripe for further research.

Disclosure statement

No potential conflict of interest was reported by the authors.

ORCID

Jingru Zhang http://orcid.org/0000-0003-4546-5576

References

Archer, M. S. (2007). *Making our way through the world: Human reflexivity and social mobility*. Cambridge, UK: Cambridge University Press.

Ateljevic, J., & Page, S. J. (2009). *Tourism and entrepreneurship: International perspectives*. London: Routledge.

Baudrillard, J. (2001). *Jean Baudrillard: Selected writings* (2 ed.). Redwood City, CA: Stanford University Press.

Beck, U., & Beck-Gernsheim, E. (2002). *Individualization: Institutionalized individualism and its social and political consequences* (Vol. 13). London: Sage.

Benson, M., & O'Reilly, K. (2009). Migration and the search for a better way of life: A critical exploration of lifestyle migration. *The Sociological Review*, *57*(4), 608–625.

Benson, M., & Osbaldiston, N. (2014). *Understanding lifestyle migration: Theoretical approaches to migration and the quest for a better way of life*. Hampshire, UK: Palgrave Macmillan.

Binkley, S. (2007). Governmentality and lifestyle studies. *Sociology Compass*, *1*(1), 111–126.

Boyatzis, R. E. (1998). *Transforming qualitative information: Thematic analysis and code development*. Thousand Oaks, CA: Sage.

Bredvold, R., & Skålén, P. (2016). Lifestyle entrepreneurs and their identity construction: A study of the tourism industry. *Tourism Management*, *56*, 96–105.

Carspecken, F. P. (1996). *Critical ethnography in educational research: A theoretical and practical guide*. London: Routledge.

Chen, L., & Yang, Z. (2014). Features of lifestyle-oriented small tourism firms and its implications. *Tourism Tribune*, *29*(8), 80–88.

Cohen, S. A., Duncan, T., & Thulemark, M. (2015). Lifestyle mobilities: The crossroads of travel, leisure and migration. *Mobilities*, *10*(1), 155–172.

Desforges, L. (2000). Traveling the world: Identity and travel biography. *Annals of Tourism Research*, *27*(4), 926–945.

Dewhurst, P., & Horobin, H. (1998). Small business owners. In R. Thomas (Ed.), *The management of small tourism and hospitality firms* (pp. 19–39). London: Cassell.

Featherstone, M. (1987). Lifestyle and consumer culture. *Theory, Culture & Society*, *4*(1), 55–70.

Gadamer, H.-G. (2004). *Truth and Method* (2 edn). New York: Continuum.

Geertz, C. (1973). *The interpretation of cultures*. New York, NY: Basic books.

Giddens, A. (1991). *Modernity and self-identity: Self and society in the late modern age*. Redwood City, CA: Stanford University Press.

Hu, X. (2013). Migratory white collars: Escape from BSG and metropolitanization trap. *China Youth Study*, (3), 32–36.

Kwong, Y. T. (1989). Naturalness and authenticity: the poetry of Tao Qian. *Chinese Literature: Essays, Articles, Reviews (CLEAR)*, *11*, 35–77.

Lefebvre, H. (1991). *Critique of everyday life* (Vol. 1). London: Verso.

Levy, S. J. (1999). *Brands, consumers, symbols and research: Sidney J Levy on marketing*. Thousand Oaks, CA: Sage Publications.

Liang, H., Xu, H., & Thomas, R. (2010). Study on the motivation and objectives of life-styled tourism businesses in Dali ancient city. *Tourism Tribune*, *25*(2), 47–53.

Löwy, M. (1987). The romantic and the marxist critique of modern civilization. *Theory and Society*, *16*(6), 891–904.

Ma, S., & Xu, H. (2016). From consumption to production: Understanding the tourism lifestyle entrepreneur in Dali. *Tourism Tribune*, *31*(5), 81–88.

MacCannell, D. (1973). Staged authenticity: Arrangements of social space in tourist settings. *American Journal of Sociology*, *79*(3), 589–603.

Marx, K. (1998). *The German ideology*. Amherst, NY: Prometheus Books.

McAdams, D. P. (1997). The case for unity in the (post) modern self. In R. D. Ashmore & L. Jussim (Eds.), *Self and identity: Fundamental issues* (pp. 46–78). New York, NY: Oxford University Press.

Notar, B. E. (2006). *Displacing desire: Travel and popular culture in China*. Honolulu: University of Hawaii Press.

O'Regan, M. (2016). A backpacker habitus: The body and dress, embodiment and the self. *Annals of Leisure Research*, *19*(3), 329–346.

O'Reilly, K. (2012). *Ethnographic methods*. London: Routledge.

Park, Y. A., Gretzel, U., & Sirakaya-Turk, E. (2007). Measuring web site quality for online travel agencies. *Journal of Travel & Tourism Marketing*, *23*(1), 15–30.

Saviolo, S., & Marazza, A. (2013). *Lifestyle brands: A guide to aspirational marketing*. Hampshire, UK: Palgrave Macmillan.

Shaw, G., & Williams, A. (2004). From lifestyle consumption to lifestyle production: Changing patterns of tourism entrepreneurship. In R. Thomas (Ed.), *Small firms in tourism: International perspectives* (pp. 99–114). Amsterdam: Elsevier Ltd.

Sobel, M. E. (1981). *Lifestyle and social structure: Concepts, definitions, analyses*. New York, NY: Academic Press.

Stone, I., & Stubbs, C. (2007). Enterprising expatriates: Lifestyle migration and entrepreneurship in rural southern Europe. *Entrepreneurship and Regional Development, 19*(5), 433–450.

Sun, X., & Xu, H. (2017). Lifestyle tourism entrepreneurs' mobility motivations: A case study on Dali and Lijiang, China. *Tourism Management Perspectives, 24*, 64–71.

Tanenbaum, J. G., Williams, A. M., Desjardins, A., & Tanenbaum, K. (2013). Democratizing technology: Pleasure, utility and expressiveness in DIY and maker practice. In W. E. Mackay, S. Brewster, & S. Bødker (Eds.), *Proceedings of the SIGCHI Conference on Human Factors in Computing Systems*, Paris, France (pp. 2603–2612). New York, NY: ACM.

Tang, X., & Xu, H. (2015). Review of lifestyle migration. *Progress in Geography, 34*(9), 1096–1106.

Thompson, J. A., & Bunderson, J. S. (2001). Work-nonwork conflict and the phenomenology of time: Beyond the balance metaphor. *Work and Occupations, 28*(1), 17–39.

Veblen, T. (1924). *Theory of the leisure class*. New York, NY: Viking Press.

Walker, E., & Brown, A. (2004). What success factors are important to small business owners? *International Small Business Journal, 22*(6), 577–594.

Wang, N. (2000). *Tourism and modernity: A sociological analysis*. Oxford, UK: Pergamon.

Wang, S., Hung, K., & Bao, J. (2014). Is lifestyle tourism business in the age of commercialization just a dream? Challenges and remedies. *Journal of China Tourism Research, 11*(1), 19–34.

Williams, A. M., & Hall, C. M. (2002). Tourism, migration, circulation and mobility. In C. M. Hall & A. M. Williams (Eds.), *Tourism and migration: New relationships between production and consumption* (pp. 1–52). Dordrecht, Netherlands: Kluwer Academic Publishers.

Williams, A. M., Shaw, G., & Greenwood, J. (1989). From tourist to tourism entrepreneur, from consumption to production: Evidence from Cornwall, England. *Environment and Planning A, 21*(12), 1639–1653.

Xu, H., Ma, S., & Jiang, L. (2017). Social interactions of lifestyle tourism entrepreneurial migrants. *Tourism Tribune, 32*(7), 69–76.

Xu, H., & Wu, Y. (2016). Lifestyle mobility in China: Context, perspective and prospects. *Mobilities, 11*(4), 509–520.

Yang, Y., Wong, K. K., & Wang, T. (2012). How do hotels choose their location? Evidence from hotels in Beijing. *International Journal of Hospitality Management, 31*(3), 675–685.

Yang, Z., & Lu, L. (2008). A preliminary study on research system and methods of tourism migration. *Geographical Research, 27*(4), 949–962.

Role Shifting Between Entrepreneur and Tourist: A Case Study on Dali and Lijiang, China

Xiaoxia Sun and Honggang Xu

ABSTRACT

This study explores how lifestyle tourism entrepreneurs manage the shift between their daily roles as entrepreneurs and tourists. Based on data collected through in-depth interviews and observations, the study produces a number of major findings. First, tourism lifestyle entrepreneurs seek to balance their personal life roles as tourists and entrepreneurs. Second, they use role identification and boundary management as strategies to manage their role shifting. Third, role conflicts are inevitable and frequent between the integrated work and life domains. This study contributes to our understanding of the challenges that lifestyle entrepreneurs face in tourism destinations from a micro role-transition perspective.

Introduction

Social life is performative and determined by the roles allocated by different social contexts (Goffman, 1959). Because work and life are the two central domains people confront on a daily basis, the transition between work and life roles forms part of daily life (Ashforth, Kreiner, & Fugate, 2000). In addition, as the economy has become increasingly competitive, the burden of work has gradually eroded the private lives of urban people, leading modern urban people to feel they have lost control over their lives (Akintayo, 2010; Guest, 2002). As a result, individuals are increasingly seeking to escape from their routine lives and activities through tourism (Chen & Chen, 2017; Gibson & Yiannakis, 2002). However, due to the financial cost, escape through tourism can only be a short-term solution. After completing their holidays, tourists must return to their normal structured and confined lives.

With the increase in mobility, a global phenomenon has been observed whereby some urban people have begun to pursue the possibility of extending their lives as tourists by becoming tourism entrepreneurs (Benson & O'Reilly, 2009; Cohen, Duncan, & Thulemark, 2015; O'Reilly & Benson, 2009; Sun & Xu, 2017; Xu & Wu, 2016). The mobility motivations and business strategies of these business owners suggests that they seek a free and self-controlled lifestyle similar to that of tourists (O'Reilly & Benson, 2009; Sun & Xu, 2017). However, tourism entrepreneurs also need to manage their businesses and perform their roles as owners and managers (Ateljevic & Doorne, 2000; Bredvold & Skålén, 2016; Marchant, 2010). Whether these entrepreneurs can successfully carry out their roles and manage their role shifting largely determines whether they can fulfil their initial mobility expectations and whether the tourist entrepreneurship is the right solution for these stressed urban people.

The research on lifestyle entrepreneurs has focused on four main areas. First, a number of studies have examined the characteristics of lifestyle entrepreneurs and the types of people who are interested in becoming lifestyle entrepreneurs (Bredvold & Skålén, 2016; Marchant, 2010; Xu & Wu, 2016). Second, research on the motivations of lifestyle entrepreneurs has shown that these individuals are clearly driven by the desire to achieve a suitable work-life balance (Ateljevic & Doorne, 2000; Benson & O'Reilly, 2009; Holmes & Zimmer, 1994; O'Reilly & Benson, 2009; Sun & Xu, 2017). Third, studies on the characteristics of lifestyle businesses have found that lifestyle entrepreneurs tend to operate home based businesses to reduce the spatial separation between work and life (Ateljevic & Doorne, 2000; Bosworth & Farrell, 2011; Carlsen, Morrison, & Weber, 2008; Marcketti, Niehm, & Fuloria, 2006; Wang, Hung, & Bao, 2015; Williams, Shaw, & Greenwood, 1989). Fourth, studies on host-guest interactions have examined the conflicts and challenges faced by lifestyle entrepreneurs (Lynch, 2005; McIntosh, Lynch, & Sweeney, 2011; Sweeney, Docherty-Hughes, & Lynch, 2018). However, few studies have investigated how lifestyle entrepreneurs manage their roles as tourists and entrepreneurs in their daily lives at a micro-level.

The role shifting perspective focuses on how individuals manage different tasks in relation to the specific social structure in which they live. Most of the studies on role shifting focus on the transition between work and family roles in formal organisational or family contexts (Ashforth et al., 2000; Burr, 1972; Desrochers, Hilton, & Larwood, 2005; Jones & Butler, 1980; Winkel & Clayton, 2010). These studies have shown that the fundamental conflicts that individuals face relate to the spatial and temporal separation between life and work in modern society. However, it is not fully understood whether this issue can be resolved by addressing the spatial and temporal separation between work and life. Tourism lifestyle entrepreneurship provides an ideal setting in which to examine this question.

This study uses a qualitative approach to examine how lifestyle entrepreneurs in Dali and Lijiang, China engage in daily role shifting between being entrepreneurs and tourists. The aim of this study is to understand the actual lives of Chinese tourism lifestyle entrepreneurs at a micro, individual level, and reveal the social structures that such lifestyle entrepreneurs face. The results of this study have important policy implications for tourism destination managers.

Literature review

Tourism lifestyle entrepreneurs

According to Bosworth and Farrell (2011), "Tourism lifestyle entrepreneurs are defined as tourism business owners who are actively pursuing a different lifestyle" (p. 1475). These entrepreneurs can either be local residents or migrants who move to a local area to start a tourism business. However, it should be noted that not every business owner is an entrepreneur, and therefore the term is used in a generic sense in this study.

Lifestyle entrepreneurs are primarily motivated by the desire to maintain a certain quality of life by earning a suitable income (Deakins & Freel, 2006; Li, Miao, Zhao, & Lehto, 2013). Because these entrepreneurs appreciate the quality of life and beautiful local environment, they are not as profit-oriented as other entrepreneurs (Ateljevic & Doorne, 2000; Carlsen et al., 2008; Lynch, 2005; Saxena, 2015; Shaw & Williams, 2004). Operating a small tourism business enables lifestyle entrepreneurs to have some degree of control over their lives, and self-employment appeals to them because it enables them to better balance their work and personal lives (Holmes & Zimmer, 1994).

Travel plays an important role in an entrepreneur's decision to start a lifestyle business. Travel seems to be a catalyst for entrepreneurs to start businesses because it opens their eyes to the different types of enterprises they can explore (Marchant, 2010). The decision to become a lifestyle entrepreneur originates from entrepreneurs' own travels as tourists, and signals the beginning of a transition period between their former lifestyle and a new lifestyle in which they seek new meaning in their lives (White & White, 2004; Williams & Hall, 2000; Williams et al., 1989). Thus, the mobility motivations of lifestyle entrepreneurs include escaping their past lives and searching for a new life (Benson & O'Reilly, 2009; Sun & Xu, 2017).

Most of the tourism lifestyle entrepreneurs in Dali and Lijiang are former tourists (Sun & Xu, 2017; Xu & Wu, 2016) who have sought to escape their urban lifestyle and have moved to the ancient town tourism destinations to pursue a more self-controlled and leisurely lifestyle in a beautiful environment (Sun & Xu, 2017; Xu & Wu, 2016). To attain this lifestyle goal, they generally set up small self-run tourism businesses to maintain a high degree of control over their work and personal lives. Operating guesthouses is the optimal choice for lifestyle migrants. In this case, the entrepreneurs rent a house from local residents, transform it into guest rooms and live there as their home (Sun & Xu, 2017; Xu & Wu, 2016). Because they live in their commercial homes, the private lives of lifestyle entrepreneurs become entangled in the commercial public spaces (Sweeney et al., 2018). Moreover, in the process of moving to the local area and starting a business, tourism lifestyle entrepreneurs act as both tourists (personal escape, pursuing a free lifestyle) and entrepreneurs (self-employed and responsible for customers).

The number and percentage of the lifestyle entrepreneurs are negatively related to the level of commercialisation of tourism destinations (Wang et al., 2015). Almost all of the lifestyle businesses in Dali and Lijiang are located in the quiet and remote parts of the ancient towns, which have relatively lower rent and are characterised by low to moderate levels of touristification (Ma, 2013; Sun & Xu, 2017). There are no official statistical data on the number and percentage of lifestyle entrepreneurs in relation to business owners. However, based on field surveys, Ma (2013) estimated that in 2012, approximately one-fifth of the businesses in the Dayan Ancient Town of Lijiang were lifestyle-led. The 249 lifestyle businesses in Dali Ancient Town in 2012 comprised about half of the businesses in the area (Ma, 2013). In contrast, in 2014, nearly 80% of the businesses in the Baisha

Ancient Town of Lijiang were lifestyle-oriented (Sun & Xu, 2017). Baisha was in the initial stages of tourism development at this time.

Roles and role shifting

Individual behaviour in social contexts is organised by and attains meaning through roles (Turner, 1990). Katz and Kahn (1978) define roles as the requirements that social systems place on individuals with respect to one's social position and the requirements associated with that position. Roles are manifest on three different levels: society (social position), culture (normative expectations) and individual (role behaviour) (Allen & Van de Vliert, 1984). Specifically, roles are derived from the expectations of others and reflect the appropriate behaviour required of particular social positions (Kahn, Wolfe, Quinn, Snoek, & Rosenthal, 1964).

Because social roles involve interacting with others, they are usually defined in terms of complementary positions, such as husband-wife, employer-employee and tourist-local resident (Allen & Van de Vliert, 1984). Thus, role shifting also reflects the social structures that define the roles. For example, tourism entrepreneurs who operate guesthouses act as entrepreneurs in their work roles, whereas their role requirements are defined by their guests and employees.

Role transition refers to the process of changing from one set of expected positional behaviours to another (Allen & Van de Vliert, 1984), "moving in and out of the roles in a social system" (Burr, 1972, p. 407) and transferring one's resources to another domain through particular behaviours (Matthews & Winkel, 2014). According to boundary theory, a role transition is a boundary-crossing movement that includes role exit and role entry (Ashforth et al., 2000). There are three types of role boundaries, namely temporal, spatial and psychological role boundaries, which are defined in terms of their permeability and flexibility (Ashforth et al., 2000; Clark, 2000). There are also two types of role transitions: macro-role transitions, which are more occasional and related to important life changes (e.g. promotion, retirement), and micro-role transitions, which occur frequently and repeatedly in daily life (e.g. the repeated transition between work and family roles) (Winkel & Clayton, 2010).

The research on role shifting has mostly been conducted in organisational contexts from an enterprise management perspective (Akintayo, 2010; Ashforth et al., 2000) and in familial contexts from a gender perspective (Burr, 1972; Desrochers et al., 2005; Jones & Butler, 1980; Winkel & Clayton, 2010). Work and life are the two central domains people confront daily, and thus a part of their lives consists of transitions between their work and life roles (Ashforth et al., 2000). Although studies of role shifting have focused on the transitions between work and the home/family domain, role shifting can also occur between work and personal life, which covers a broad range of activities, such as leisure, travel and hobbies (Keeney, Boyd, Sinha, Westring, & Ryan, 2013; Li et al., 2013). This study explores the daily micro-level transitions between tourism entrepreneurs' work and personal lives.

Role transition and role conflict

A single position may have several role senders, but the role expectations of different role senders may be contradictory and bring about intra-role conflict (Kahn et al., 1964). Furthermore, because individuals usually occupy multiple roles at any one time in their daily lives, such as their gender, work and family roles, they receive role expectations from multiple role senders (Allen & Van de Vliert, 1984).

Role conflict is defined as the concurrent appearance of multiple incompatible role expectations (Coverman, 1989; Stryker & Macke, 1978), and is related to "the extent to which a person experiences pressures within one role that are incompatible with the pressures that arise within another role" (Kopelman, Greenhaus, & Connolly, 1983, p. 201). The increased transitions between roles can lead to increased role conflict (Desrochers et al., 2005; Matthews, Barnes-Farrell, & Bulger, 2010; Matthews & Winkel, 2014). Most of the studies on role conflict have been conducted in an organisational perspective and have found that role conflict is associated with organisational stress (Lin, Huang, Yang, & Chiang, 2014; Lin, Wong, & Ho, 2015; Michel, Mitchelson, Kotrba, LeBreton, & Baltes, 2009; Schmidt, Roesler, Kusserow, & Rau, 2014; Stryker & Macke, 1978). Work-family role conflicts can have negative work-related (job satisfaction, organisational commitment), non-work-related (life/leisure/family satisfaction) and stress-related (psychological strain, depression, burnout, work/life-related stress) effects (Akintayo, 2010; Allen, Herst, Bruck, & Sutton, 2000; Kossek, Noe, & Demarr, 1999; Michel et al., 2009; Schmidt et al., 2014).

According to the boundary flexibility and permeability perspectives, roles can be divided into a continuum of segmented and integrated roles (Ashforth et al., 2000). For example, because lifestyle-oriented B&B innkeepers are usually self-employed, live on the property and provide personalised services at home, the spatial and temporal boundaries between their work and personal lives tend to overlap (Hsieh, 2010; Hsieh & Lin, 2010). Frequent and unpredictable role transitions can occur between the overlapping domains, and role blurring can bring about conflict and interruptions (Ashforth et al., 2000; Winkel & Clayton, 2010). Small home-based business owners are attracted to the flexible lifestyle and motivated by the possibility of achieving work-family balance. However, this may not be the case for female entrepreneurs who have been found to face role conflicts between their work and family lives (Agarwal & Lenka, 2015; Rehman & Azam Roomi, 2012; Shelton, 2006; Walker, Wang, & Redmond, 2008).

Thus, more research is needed on the role conflicts experienced by tourism entrepreneurs. In addition to the role conflicts between work and family from a gender perspective, there may be other role conflicts that affect tourism lifestyle entrepreneurs. Although it has been suggested that tourism entrepreneurs' business motivations reflect their desire to be tourists, little research has examined their roles as tourists and entrepreneurs. This study is also a response to the call for more qualitative research on the actual lives of tourism lifestyle entrepreneurs (Hsieh, 2010).

Methodology

This study follows the research paradigm of interpretivism. Qualitative research method is used to examine the lives of tourism lifestyle entrepreneurs in Dali and Lijiang. Semi-structured in-depth interviews and participant observations were used to collect the data. The interviews examined the participants' micro-role transitions and the strategies they used to maintain their work-life balance (Ashforth et al., 2000; Hsieh, 2010). Although the researcher made efforts to conduct the interview without influencing the interviewees, it is also impossible to avoid that the interviewees may provide distorted information particularly when the interviews were recorded on the spot (Bosworth & Farrell, 2011; Polkinghorne, 2007). Therefore, the first author also conducted participant observations to objectively record how entrepreneurs' interact with other people.

Inductive reasoning strategies were used in data analysis and reporting (Petty, Thomson, & Stew, 2012). In this process, it was aware that subjectivity of researchers plays an important role in understanding these data. The researchers therefore tended to keep on reflecting and presenting how they arrived at these conclusions from the data.

1) Dali and Lijiang

Dali and Lijiang, which are located in north-western Yunnan Province in China, are famous and popular tourist destinations that have attracted many lifestyle entrepreneur migrants from the big cities (Sun & Xu, 2017; Xu & Ma, 2012; Xu & Wu, 2016). Dali is a National Historical and Cultural City and was one of the first opening-up cities in China. Since 1984, the city has attracted many foreign tourists, and after 1996, a number of bars, Western restaurants and cafés began to appear on "Foreigner Street" in Dali Ancient Town. Attracted by the slow pace of life, increasing numbers of domestic migrants have begun to move to Dali from the big cities to set up small businesses and start a new life (Ma, 2013). Nearly 38.59 million tourists visited Dali in 2016 (Tourism Development Committee of Dali). Lijiang Ancient Town (which includes Dayan Ancient Town, Shuhe Ancient Town and Baisha Ancient Town) was included in UNESCO's World Cultural Heritage List in 1997. Since then Lijiang has become a famous tourism destination. Attracted by the ancient towns and the local Naxi culture, the number of tourists visiting Lijiang has steadily increased, reaching 40.69 million in 2017 (Tourism Development Committee of Lijiang). The natural landscape and slow pace of life in Lijiang have also attracted numerous lifestyle migrants (Sun & Xu, 2017).

2) Data collection

The fieldwork for this study was conducted by the first author in Dali and Lijiang in September and October of 2014. To identify the lifestyle entrepreneurs, before she went to the field, the first author contacted a few B&B entrepreneurs introduced through Ma, who studied lifestyle entrepreneurs in Dali in 2013. When she arrived in Dali and Lijiang, the first author used a snowball sampling strategy to select the participants for the study. The lifestyle entrepreneurs have good social network. So, after one is interviewed, he or she would introduce his or her friends to the researcher. They even take her to the meeting places where their social activities are taken place. Therefore the first author can easily obtain opportunities to talk to potential interviewees and if they agree, appointments for formal interviews can be made. In addition, she also attempted to identify lifestyle entrepreneurs herself by observing their businesses and talking to the owners and staff. The design, layout and decoration of the lifestyle entrepreneurs' businesses were identified and the first author engaged in a simple dialogue with the owners on their motivations to move to the area and start a business. These two steps revealed that the entrepreneurs were more lifestyle-oriented than profit-oriented, and enabled suitable participants to be selected for further study. The first author made efforts to kept value free and not to influence interviewees during the data collection process, and only played the role of promoting the process of talking using the exploratory questions (Bosworth & Farrell, 2011).

Twenty-one lifestyle entrepreneurs agreed to participate in face-to-face semi-structured interviews in their guesthouses. The respondents were labelled by acronyms representing their place of residence (Lijiang-LJ or Dali-DL) and numbers (from 1 to 21), such as LJ1 (see Table 1). The profiles of the interviewed lifestyle entrepreneurs are presented in Table 1. Fourteen respondents resided in Lijiang, of which six lived in Dayan Ancient Town and eight in Baisha Ancient Town. The other seven respondents resided in Dali Ancient Town. Of the 21 respondents, 13 were male and eight were female. Nearly half of the entrepreneurs interviewed were between 30 and 39 years old and 16 respondents had a college education or above, indicating high levels of education. Three-quarters of the respondents had been living in Dali/Lijiang for less than three years.

Table 1. The profiles and related survey information of the respondents.

Label of respondents	Gender	Age	Education level	Marital status	Origin	Duration in Lijiang/ Dali	Operating pattern	Place	Interview time (minutes)	Observation or not	Days the first author lived in that guesthouse
LJ1	Male	40	College	Married	Henan Province	1.5 years	On his own	Dayan Ancient Town in Lijiang	40		
LJ2	Female	42	College	Married	Liaoning Province	5 years	Partnership with spouse	Dayan Ancient Town in Lijiang	61		
LJ3	Female	33	College	Married	Guangdong Province	4 years	Partnership with spouse	Dayan Ancient Town in Lijiang	113		
LJ4	Female	37	College	Single	Fujian Province	3 years	On her own	Dayan Ancient Town in Lijiang	57	Yes	1 day
LJ5	Male	32	College	Married	Guangdong Province	6 years	Partnership with spouse	Dayan Ancient Town in Lijiang	65	Yes	1 day
LJ6	Male	40	Junior college	Married	Guangdong Province	7 years	On his own	Dayan Ancient Town in Lijiang	58	Yes	1 day
LJ7	Female	34	Junior college	Divorced	Guangdong Province	4 months	On her own	Baisha Ancient Town in Lijiang	64		
LJ8	Female	33	College	Married	Zhejiang Province	1 year	Partnership with spouse	Baisha Ancient Town in Lijiang	43		
LJ9	Male	50s	College	Married	Beijing	2 years	Partnership with friend	Baisha Ancient Town in Lijiang	42		
LJ10	Male	41	Junior college	Single	Shanghai	2 years	Partnership with friend	Baisha Ancient Town in Lijiang	51		
LJ11	Female	33	Junior college	Divorced	Chongqing	2 years	On her own	Baisha Ancient Town in Lijiang	119		
LJ12	Female	26	Master	Single	Hunan Province	1.5 years	Partnership with friends	Baisha Ancient Town in Lijiang	83		
LJ13	Female	40s	College	Married	Yunnan Province	1 year	Partnership with spouse	Baisha Ancient Town in Lijiang	58		
LJ14	Male	50s	College	Married	Guangdong Province	1 year	Partnership with spouse	Baisha Ancient Town in Lijiang	51		
DL15	Male	23	College	Single	Hubei Province	5 months	Partnership with girlfriend	Dali Ancient Town	36	Yes	1 day
DL16	Male	27	College	Single	Chongqing	1 year	Partnership with friend	Dali Ancient Town	82		
DL17	Male	32	College	Single	Guangdong Province	1.5 years	On his own	Dali Ancient Town	36	Yes	1 day
DL18	Male	32	Junior college	Single	Jiangsu Province	1 year	Partnership with parents	Dali Ancient Town	43	Yes	2 days
DL19	Male	28	College	Single	Guangdong Province	1.5 years	Partnership with girlfriend	Dali Ancient Town	48		
DL20	Male	31	College	Married	Guangdong Province	2 years	Partnership with spouse	Dali Ancient Town	68	Yes	2 days
DL21	Male	36	College	Single	Beijing	2.5 years	Partnership with partner	Dali Ancient Town	85		

Five of the single entrepreneurs in Dali were in good relationships, and subsequently became married. Seventy percent of the lifestyle entrepreneurs operated guesthouses in partnership with friends, lovers or family members. All of the respondents lived on their properties except one.

Each entrepreneur was interviewed once, and the interview times ranged between half an hour and two hours. The interview questions focused on the entrepreneurs' migration stories (their origin, work and living situation before moving to the tourism destination, tourism experience in Dali/Lijiang, impression of the tourism destination before moving, reasons for leaving their place of origin and moving to the tourism destination, migration time and desired lifestyle), their entrepreneurship motivations (reasons for starting a business, choice of business type and local house, design of the guesthouse, rent of the house, length of time they prepared for the business opening, operating style of the business, number of employees) and their strategies for coping with their daily work and personal life activities at the tourism destination (daily work time arrangements, daily leisure time activities, receiving customers, hiring employees, dealing with their employees, relationships with friends and their social circle in the tourism destination, relationship with local people and difficulties in managing the business). The interviews also included questions on the entrepreneurs' psychological perception of their work and personal lives (the satisfying and dissatisfying facets of their businesses and lives in the tourism destination and their physical and psychological satisfaction) and their future plans (the length of time they plan to live in Dali/Lijiang, their future plans for their businesses and personal lives). The interviews largely followed the outline of the prepared interview questions, although the questions were varied based on the interviewees' on-the-spot responses. Because the first author and the respondents were all Chinese, all of the interviews were conducted in Chinese for ease of communication. All of the interviews were recorded using a digital voice recorder, with the respondents' permission. When the fieldwork was completed, the interview records were transcribed into Chinese text, around 103,800 words, and the quotes used in this study were translated into English by the researchers.

Participant observations were also carried out at seven guesthouses in which the first author stayed as a guest for one or two days each (see Table 1). The first author checked in at the guesthouses whose owners were identified as lifestyle entrepreneurs, and she conducted observations during her stay in a relatively unstructured manner. After receiving the participants' consent, the observations focused on the entrepreneurs' daily activities. To obtain detailed observational information, as a guest, the first author joined in the daily leisure activities at the guesthouses, such as eating with the owners, employees and other guests, chatting, drinking tea and watching TV. Moreover, at other times during the participant observations, the researcher usually stayed in public spaces, such as the courtyard or reception hall, to observe the entrepreneurs' daily activities, which were documented in writing by the researcher. Photos were also taken with the permission of the respondents. The researchers also analysed field notes of observations to verify the interview information and derive conclusions on role shift strategies. Using triangulation by combining in-depth interview and participant observation, this study tried to improve the reliability and validity.

3) Data analysis

The data were analysed using a thematic analysis approach. Thematic analysis is used to identify, analyse and describe the themes or patterns within qualitative datasets (Braun & Clarke, 2006). The thematic analysis comprised six phases: gaining familiarity with the data, producing initial codes, searching for themes, reviewing themes, defining and naming themes and generating the report (Braun & Clarke, 2006). First, after the data collection and transcriptions were completed, the transcripts of the interviews and the observation notes were repeatedly read to gain familiarity with the data. Second, to reduce the amount of text and obtain the key meanings, open-coding was carried out by the two authors with a focus on the research questions. The coding results were then compared and the themes were extracted and collated. Finally, the two authors reached consensus on the names of the themes, which were as follows: conflict, role identification and tourism commercialisation.

The three parts of the results section in this study correspond to the following research questions: migration goals of the lifestyle entrepreneurs and their motivations for starting their businesses, strategies for coping with their work and life roles, and the work and life conflicts encountered as a tourism entrepreneur.

Results

Lifestyle entrepreneurs escape the big cities for a free and leisurely life

Most of the lifestyle tourism entrepreneurs in Dali and Lijiang evolved into entrepreneurs through their activities as tourists and through their search for a free lifestyle similar to that associated with tourists (Sun & Xu,

2017; Xu & Wu, 2016). Like tourists, they had migrated to escape from work, personal setbacks, relationships and the negative elements of big-city living (Benson & O'Reilly, 2009; Sun & Xu, 2017), and their tourism experiences in Dali/Lijiang opened their eyes to the possibility of a different lifestyle.

> *When I visited Lijiang as a tourist, I observed the enviable lifestyle of a guesthouse owner who was my former colleague. She was leisurely every day and was able to travel to many places. I admired her life so much that I resigned after going back home and moved to Lijiang and started a guesthouse to lead a high-quality life and to have enough leisure time to do things I liked, such as drawing, travelling and shopping.* **DL16**

In this quote, the young tourism entrepreneur (**DL16**) describes the process of role-shifting from a tourist to an entrepreneur in a tourism destination. Two years later, he moved from Lijiang to Dali because of the cheaper rent in Dali. From a tourist's perspective, this small tourism business owner lives like a tourist and can easily and freely shift roles between that of an entrepreneur and that of a tourist.

All of the respondents had experienced living in big cities and being tourists in Dali/Lijiang. The lifestyle migrants moved to Dali/Lijiang to have the time to do what they were interested in, such as reading books, drawing and travelling (**DL21, DL16**). Because they were their own boss and self-employed, the felt they had high levels of autonomy and flexibility in their work and personal lives. Operating a guesthouse was an optimal choice for migrants with sufficient start-up capital, because living in a local house with a yard satisfied their dream of an idyllic life (**LJ3, LJ5, LJ13, LJ14, DL19, DL21**). However, the younger migrants who did not have sufficient financial resources tended to share the economic burden with friends (**LJ12, DL16, DL21**).

> *We want to live an idyllic life in the quiet ancient town, and plant vegetables and flowers in the yard.* **LJ13**

> *I feel that the living environment here is a bit like that of my childhood, and the guesthouse can satisfy my dream of having a courtyard.* **LJ11**

> *Freedom of time may be the greatest temptation of operating an inn, it means that you are able to arrange your time flexibly and earn money to live. You can leave for travel or home in the tourist off-season, but if you are an office worker, you do not have the time for leisure.* **LJ8**

A quiet and peaceful position and low rent were the key considerations for choosing where to start a B&B (**LJ8, LJ10, LJ13, DL18, DL19, DL21**). Thus, Baisha Ancient Town and Dali were more popular with the new lifestyle migrants than Dayan Ancient Town in Lijiang. The entrepreneurs designed the rooms and the public and private spaces according to their personal preferences. Based on the first author's observations, the owners usually planted flowers and vegetables in the yard, placed a rocking chair in the yard or on the roof for sunbathing (see Figure 1), and placed a tea table in the hall for friends, customers and family members. Some of the entrepreneurs even decorated the public spaces with books, put up screens for showing films (see Figure 1) and offered their guests TVs, ping pong tables and other such amenities (**LJ4, LJ5, LJ12, DL15, DL20, DL21**).

Most of the lifestyle migrants found that they did not require professional knowledge to run their businesses and that they could easily operate their guesthouses by making efforts in network promotion and employing someone to do the cleaning (**LJ1, LJ2**).

> *It's not difficult to run a guesthouse, as long as you do a good job of internet marketing, and you can manage the guesthouse easily after the customer base has become relatively stable.* **LJ1**

Strategies to cope with role shifting

Different from the employees in formal organisations, lifestyle entrepreneurs have flexible and autonomous lives (Hsieh & Lin, 2010), and can shift roles between being entrepreneurs and tourists. The capacity to shift from one domain to another can lessen the likelihood of role conflict (Ashforth et al., 2000). However, a study found that home workers prefer to work at specified times and places to set up a work-home boundary (Clark, 2000). Moreover, because B&B operators' work and personal lives are limited to the same location and the same group of clients (Li et al., 2013), it is difficult for them to develop physical, temporal and psychological borders (Hsieh, 2010). Thus, the strategies for coping with role conflict and shifting, such as role identification and boundary management, are very important for lifestyle entrepreneurs.

These individuals can smoothly shift between the roles of entrepreneur and tourist by adapting the role of entrepreneur to integrate the entrepreneur and tourist roles. During interactions with tourists, guesthouse owners often shift between the roles of host and businessperson according to their needs (Zhang, Xu, & Xing, 2017). The first author stayed in **LJ4**'s guesthouse on a mid-autumn day during a big reunion festival in China. An advertisement had been posted at the door announcing that guests could dine together in the evening, and the first author asked to join the dinner party and the owner agreed. The entrepreneur was observed buying food

Figure 1. The public space of guesthouses.
Source: the first author's photos (Aug, 2014).

with two local friends in the afternoon and then preparing the dinner together with the two friends. Eight people attended the dinner, including the owner, an employee, two local friends and four guests (including the first author) staying at the guesthouse (see the left picture in Figure 2). We had a nice evening, eating, drinking tea and singing songs together accompanied by a friend who played guitar (see the right picture in Figure 2). After dinner, the first author offered to wash the dishes, but the entrepreneur stopped her and asked her to give good comments on Qunar. The owner happily participated in the dinner with her friends and guests just like a tourist, and played the role of entrepreneur when she asked the guests to give positive comments on the booking website. As a lifestyle entrepreneur, the owner, **LJ4**, was able

Figure 2. The dinner party in the guesthouse of LJ4.
Source: the first author's photos (Sep, 2014)

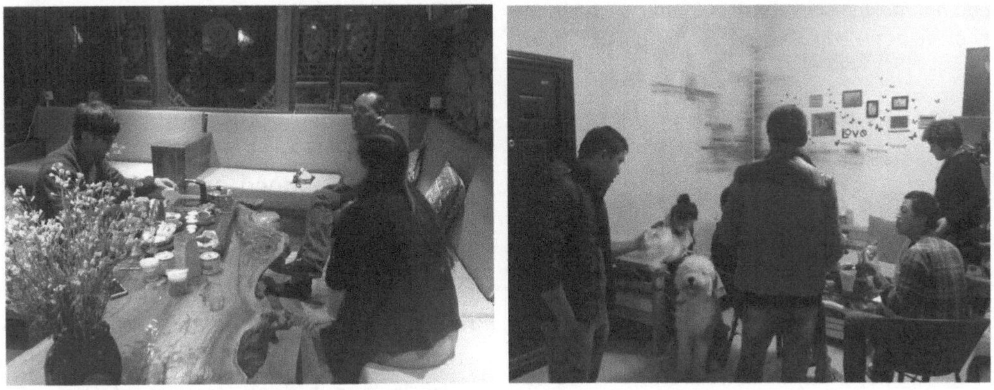

Figure 3. The lifestyle entrepreneurs were drinking tea and chatting in the guesthouses.
Source: the first author's photos (Aug, 2014)

to smoothly shift between the roles of tourist and entrepreneur.

By combining their personal interests with their work the owners were able to develop a strong role identification as lifestyle entrepreneurs and enjoy the process of role integration. In the youth hostels, the business owners and visitors often watched movies, played games and chatted together in the public spaces (**LJ4, LJ7, LJ10, DL15, DL21**). Generally, these activities were based on the owners' hobbies, and thus they were able to seamlessly combine their work and personal lives. **DL21** did not consider chatting with guests to be work. Rather, the interaction made him feel very happy and he enjoyed getting to know different people and making friends. In his life role as a tourist, **DL21** enjoyed making new friends with his guests and keeping in touch with them, which he felt enriched his life, whereas, as an entrepreneur, he hoped that his guests would become returning customers or even voluntary workers. Many of his guests recommended the guesthouse to friends, which was good for his business.

Boundary management is an important factor in successful role transition (Winkel & Clayton, 2010) and entrepreneurs can use temporal strategies to set up role boundaries to reduce conflict. Most of the lifestyle entrepreneurs in this study did not have fixed working hours, and hiring employees, such as front desk staff and cleaners, freed the entrepreneurs from having to work long hours and complete diverse tasks (**LJ4, LJ5**). Setting a gate password or using swiped door cards helped the entrepreneurs to reduce the service time, as these measures enabled latecomers to enter the guesthouses by themselves (**LJ4, LJ6, DL19**). If guests wished to leave early in the morning, they could check out the night before based on mutual trust and let themselves out in the morning (**LJ10, LJ12, DL20**).

> *I employ a guesthouse manager for reception. We usually close the door at 11 p.m., and the guests who come back later can swipe the door card to get in. We open the guesthouse at 8–9 a.m., and the guests who leave earlier can open the door by themselves.* **LJ4**

> *The opening time of our guesthouse is not fixed, and it depends on the situation. Our guests who leave early in the morning check out the night before.* **DL20**

Lifestyle entrepreneurs can create leisure time through successful temporal boundary management. Thus, lifestyle entrepreneurs with sufficient leisure time can live a life as free as a tourist. The lifestyle entrepreneurs in this study had frequent contact and maintained good relationships with other entrepreneurs, often making their rounds in the daytime, having dinner together, drinking tea and chatting in the afternoon or evening (see the left picture in Figure 3) (**LJ7, LJ8, LJ11, DL20**). In the right picture in Figure 3, one night in DL20's guesthouse, several owners were tasting hamburgers from a restaurant opened by a new comer of lifestyle entrepreneur, who seek advice to improve the tastes. During the tourist off-season, the lifestyle entrepreneurs could easily adopt the role of tourist based on the vacation schedule. Nineteen of the respondents closed their guesthouses or found other people (parents, friends or employees) to manage the hostel, and left for other tourism destinations (all except **LJ2, DL17**).

> *We live a free life except during the busy season (summer vacation, National Day holiday and Spring Festival holiday). We live leisurely lives in the tourism-off season, when a lot of owners leave for travel.* **LJ10**

Marcketti et al. (2006) found that although many business owners were not satisfied with their income, their quality of life and satisfaction with their business had been improved through lifestyle entrepreneurship.

Although we do not make much money in Lijiang, we have a high quality of life as we can often go out for fun. **LJ3**

The inevitable conflict between work and personal life

Lifestyle entrepreneurs experience a transition in moving from the familiar role of a tourist to the relatively unfamiliar role of a tourism entrepreneur (Duchscher, 2009). In this study, the entrepreneurs found the switch from tourist to entrepreneur was more difficult than they had expected. The lifestyle entrepreneurs had to spend much time and energy in setting up their businesses. For example, **LJ14** took one year to rebuild and design his house in Baisha Ancient Town, which he found very tiring. **LJ12** and **LJ15** worked hard in other guesthouses for half a year to learn the business before starting their guesthouses.

I am exhausted because the house in the countryside is different from the one in the city. Employing a local construction team to tear down old houses and build new ones, I had to communicate with contractors on the spot and adjust the design scheme quickly. I designed the inn's layout and the furnishings of every room, and bought all relevant items inside the room. The preparations for the opening of the guesthouse were very tiring. **LJ13**

Role blurring increases the mutual effects of interruption (Burke, 1991). Lifestyle-oriented B&B entrepreneurs can experience three types of conflicts between their work and personal lives: temporal conflicts, spatial conflicts and mental conflicts. Temporal conflicts occur when someone has to spend so much time on one role that they are unable to fulfil other roles (Hsieh, 2010). Operating a guesthouse was time-consuming work for many of the entrepreneurs in this study, as they had to wait for their guests to check in and out, and the guests could request services at any time. However, the guesthouse owners' working hours were not particularly fixed. During the peak season, some of the owners needed to get up before 7 a.m. to receive newly arrived guests and had to wait at night until all of the guests returned (**LJ2, LJ3, DL15, DL16, DL20**). The entrepreneurs also needed to stay at the guesthouse to address the various needs of the guests, such as loss of power, problems with the electricity, clogged toilets and other problems reported by their guests (**LJ2, LJ6, LJ11, DL15, DL16, DL20**). **DL16** found he was stuck at work and could not easily live freely as a tourist.

I thought I should have enough time for travelling and shopping before migration. In fact, I invested so much time in the business that I had no time to do these things. **DL16**

Spatial conflicts between work and personal life stem from the different demands of the private and commercial spaces. Many entrepreneurs find it difficult to divide their homes into private and commercial spaces (Sweeney et al., 2018). In this study, the B&B innkeepers designed private spaces for their personal lives and hobbies and public spaces for their family members and guests (**LJ6, LJ10, LJ13, LJ14, DL15, DL21**). When spatial conflict occurred, the entrepreneurs abandoned their personal tasks to satisfy their customers (**LJ3, LJ6, DL15, DL21**). **LJ6** kept nine large dogs in two hidden corners of the yard so that the guests could not see them from the guesthouse. Three dogs were in the north-western corner, and the other six dogs were in the south-eastern corner, which was a separate small courtyard. The dogs lived on the first floor and the owner lived on the second floor. Although the owner loved the dogs, he only let the dogs go out in the yard when there were no guests present, in consideration of the guests' fear of dogs. Similarly, the entrepreneurs who had children experienced spatial conflict between their family life and business work. For example, **LJ5** stated that he chose not to live in the guesthouse.

Our daily schedule is based on our guests, we can answer the phone and open the door for them at two o'clock in the morning. But since having a child, we need the regularity of life and can't accept these disturbances. Furthermore, a child's crying can also affect the guests. So we moved out of the guesthouse and rented another yard to live in for the child's health and the B&B's operation. **LJ5**

Mental conflicts stemmed from the high contrast between the personal needs of the entrepreneurs and their guests' demands. Working in a service industry is demanding, and the lifestyle entrepreneurs needed to provide face-to-face services for various customers with different personalities and service requirements, which some of the respondents found to be very annoying (**LJ2, LJ3**). On one occasion, **LJ3** had to ask several customers to leave at 2 a.m. because they were so noisy that they disturbed the owner and other customers. Although they found it difficult, entrepreneurs had to serve various types of people, including those they disliked and did not want to talk to (**DL16**). **DL21** ran a youth hostel, but was gradually growing weary of operating the inn, as he said:

I feel tired and vexed especially when there are so many guests; my head is buzzing. Particularly, during spring festival, it is the busiest time to serve guests, whereas prior to this family members could dine together and watch TV quietly. **DL21**

Discussion

China is undergoing significant social, economic and cultural changes. As China undergoes this rapid transition toward modernity, increasing numbers of people are seeking a better quality of life by improving their work-life balance (Sun & Xu, 2017; Xu & Wu, 2016). The tourism industry and tourism destinations are perceived to provide such opportunities. Tourism destinations that are rich in natural and cultural resources and which appear to be relaxed and easy places tend to attract lifestyle entrepreneurs (Benson & O'Reilly, 2009; Xu & Ma, 2012).

The relaxed and leisurely atmosphere in ancient town tourism destinations enables outside lifestyle entrepreneurs to feel like tourists. As special tourism businesses, guest houses also allow entrepreneurs to control the rhythm of their working lives to some extent. Guest houses provide the possibility of integrating life and work temporally and spatially and thus avoiding the stress of splitting life and work. Moreover, the guests can often be treated as friends and therefore serving them is not only a work assignment but also a source of pleasure. Thus, tourism entrepreneurs can achieve their goal of enjoying the life of a tourist while managing a business.

Through careful examination of the role shifting and role management of tourism entrepreneurs, it can be seen that tourism entrepreneurs constantly face the challenge of role conflict and that integrating the roles of entrepreneur and tourist is not as easy as imagined. The conflicts are due to individual factors affecting lifestyle entrepreneurs, the social structure (Ashforth et al., 2000; Meleis, Sawyer, Im, Hilfinger Messias, & Schumacher, 2000) and the Chinese context.

Individual factors

From the owners' personal perspectives, individual financial muscle and business capacity determined how much pressure they felt from their businesses, and further influenced the role conflicts between entrepreneur and tourist. The migrant entrepreneurs had to invest significant resources in setting up and operating their guesthouses and their businesses were based on a long return cycle. The entrepreneurs spent considerable amounts of money on rebuilding their houses or decorating rooms and had to pay rent for several years at a time. The middle-aged and retired entrepreneurs with a strong financial base were less concerned about their businesses (**LJ1, LJ9, LJ10, LJ13, LJ14**) and enjoyed their leisure time more. In contrast, younger owners without sufficient savings had to struggle to balance their returns and spending (**LJ11, LJ12, DL15, DL17**). Sometimes the entrepreneurs were trapped in a vicious cycle. For example, **DL17** did not have enough money to upgrade the hostel to raise the rate. At the end of 2014, he was under enormous financial pressure to pay the rent (USD38,170/CNY240,000) for the next three years. To compensate, he sought to broaden his income channels by opening a bar on the first floor of the guesthouse and establishing sales channels for a milk factory. However, in devoting so much energy to making money, **DL17** felt too exhausted to enjoy his personal life and partially attributed his situation to poor management.

> *I do not have enough money to invest in the rooms, and I also do not have enough time to do this. I do not have the ability to operate the inn well.* **DL17**

The owners who had a strong business acumen were able to adapt their management strategies to reduce the conflict between work and life, such as by employing reception supervisors, volunteers and cleaners (**LJ1, LJ4, LJ5, LJ9, LJ13, DL21**). **LJ1**, in his 40s, had worked for 16 years and accumulated a large amount of savings and abundant experience in enterprise management before moving to Lijiang. Thus, it was easy for him to manage a small tourism business. For him, good marketing through the Internet and maintenance of his customer base were enough to operate his guesthouse successfully. Moreover, he employed a woman to receive customers and do the cleaning so that he could sleep more and enjoy travelling with local friends during the off-seasons.

Social structure

Lifestyle entrepreneurs are embedded in their social structure (Bredvold & Skålén, 2016). In view of the external environmental factors, the level of tourism commercialisation in a tourism destination is directly proportionate to the housing rents, living costs and the level of business competition (Wang et al., 2015). In areas with rising housing rents and living costs, entrepreneurs need to spend more time on their businesses and hence have less leisure time. In contrast, the owners in areas with low rents and low living costs can devote more time to leisure and more easily shift between work and life. For example, in 2011, **DL21** took out a lease on a courtyard from a local resident for 20 years. To ensure the rent remained the same at USD7,952 (CNY50,000) per year for the first 10 years, **DL21** paid the landlord the first eight years' rent at one time. This relatively low level of rent meant that **DL21** could easily balance work and life. However, with the rising tourism commercialisation in Dali and Lijiang, more and more local residents have begun to rent their houses to migrants, thus leading to rising house rents. In 2013, the annual house rent in Baisha Ancient Town in Lijiang was USD11,945 (CNY75,000). The following year the rent had increased to USD15,131 (CNY95,000), a 26.7% annual increase (**LJ12**). As a result, some early migrant entrepreneurs began to sublease their guesthouses at considerably higher prices to new migrants. The sublease fees included one-time transfer expenses and the annual rent. Many advertisements for guesthouses were observed pasted on the walls in the ancient towns of Dayan and Dali, with the transfer fees ranging from USD159,000 (CNY1,000,000) to USD477,000 (CNY3,000,000), and annual house rents ranging from USD15,900 (CNY100,000) to USD31,800 (CNY200,000). In 2014, **DL15** and his girlfriend leased a youth hostel run by a former entrepreneur migrant for three years at an annual rent of USD15,900 (CNY100,000). The B&B operators had to expend much more energy and wisdom in Internet marketing and in setting up reception services to maintain their incomes. Some guesthouse entrepreneurs even began selling local specialties through social media (such as brown sugar, apples, tea, medical herbs, jade and mushrooms) to increase their income. Some migrant entrepreneurs chose to leave the tourism destination due to the economic pressure (**LJ5, LJ6, LJ7, DL21**).

In addition to rising costs, the change in the customer base required the lifestyle entrepreneurs to abandon their flexible arrangements with their customers and employees. With the increasing infrastructure, mass tourists have gradually replaced the former individual backpacker-type tourists, and the relationship between the entrepreneurs and their customers has changed from host-guest to service-provider and service-consumer. When the visitors were guests, they were generally interested in the lifestyle entrepreneurs' lives and were willing to make friends with the owners. Hence, the guests were willing to communicate with the owners and even participate in the owners' daily lives, such as participating in cooking. In their life role of tourists, the owners were happy to develop relationships with their guests. However, when the visitors became customers requesting more services, the work roles of the entrepreneurs shifted, as the customers became more demanding and the entrepreneurs found the work tiring (**LJ2, DL16**). **LJ2** said that she noticed the changes in the tourist market.

> *Several years ago, we often organised activities with tourists, such as chatting and barbecuing. All of us went to buy food together and participated in the whole process during the barbecues, making everyone very happy. However, now, if we invite customers to have a barbecue, they will come and wait to eat when we have prepared the food and then leave directly after the meal. It is better for us to have a rest rather than organise activities without a friendly atmosphere.***LJ2**

Chinese context

The role shifting conflicts of lifestyle entrepreneurs also reflect the special Chinese context. For instance, many young people were found to have made the decision to be tourism entrepreneurs very rapidly and without careful planning (Sun & Xu, 2017). This could be due to the current entrepreneurship culture in China. Indeed, over the past 40 years, many people have successfully established small businesses (Xu & Ma, 2012). In addition, tourism is considered to be an emerging industry in China and thus to provide good business opportunities.

The Hukou household registration system, which controls the voluntary migration between different places in China (Lu, 2008), is another social factor that entrepreneurs need to consider before moving to a tourism destination (Xu & Wu, 2016). Although mobility has greatly improved in China, Hukou still serves as a barrier to free mobility. Importantly, tourism entrepreneurs who migrate to tourism towns remain tied to their Hukou in the big cities. As a result, many of the respondents in this study had short plans for their lives in the tourism destinations, with the change serving as a transition or a rest before entering another phase of life (**LJ1, LJ7, LJ11, DL15, DL16, DL18**). This factor may have also reduced the amount of effort they devoted to planning and preparing their businesses.

Conclusion

China has undergone great changes over the past 40 years, transforming from a sedentary society to a highly mobile one. A variety of mobility patterns can now be observed, with more and more people seeking to develop their lifestyle rather than simply working to survive. Tourism lifestyle entrepreneurs are such an emerging group (Benson & O'Reilly, 2009; Cohen et al., 2015; O'Reilly & Benson, 2009; Xu & Wu, 2016). Although their numbers are still limited, lifestyle entrepreneurs represent a new social trend of individuals seeking to escape from the structured urban life to a freestyle tourist-like life through operating a small business in a tourism destination. However, this study shows that although the remote tourism destinations can temporarily help these entrepreneurs to balance their leisure and work lives, the increasing commercialisation of the destinations has the potential to undermine the conditions that have enabled them to balance their roles as tourists and entrepreneurs. In particular, the increasing rents, standardised service quality requirements of mass tourists and rising labour costs have made it increasingly difficult for tourism entrepreneurs to run their businesses. With the modern commercial supplier-customer model gradually replacing the host-guest model, tourism entrepreneurs are finding it increasingly difficult to shift between their life and work roles, resulting in increasing role conflicts, especially for those who do not have sufficient business experience.

This study makes a number of contributions to the literature. First, although lifestyle entrepreneurs have received some scholarly interest, a micro-level examination of their daily practices is lacking. By conducting a micro-level analysis of the roles and role shifting of tourism operators in Lijiang and Dali, this study has revealed the daily practices of tourism lifestyle entrepreneur. Specifically, tourism entrepreneurs tend to use role identification and boundary management to balance their daily work and lives. However, although their tourism businesses have enabled the lifestyle entrepreneurs to blur the boundaries between their roles as tourists and entrepreneurs, the socio-economic structure of the tourism industry and tourism destinations remains relatively rigid and does not leave them much room to easily shift between the two roles.

Second, this study contributes to the research on the role shifting between work and life. By examining the integration between the respondents' work and personal life roles, this study shows that role conflicts are inevitable, which is in accordance with previous studies (Ashforth et al., 2000; Rehman & Azam Roomi, 2012; Shelton, 2006; Walker et al., 2008). However, this study further reveals that the spatial integration of work and life in guesthouses cannot solve this problem. Although individuals have different capacities in managing their role shifting, the broader social structure of the destinations and the tourism industry does not provide a good environment for achieving a suitable work-life balance. Rather than the spatial separation of work and home, the increasing pressure on tourism entrepreneurs stems from the increasingly commodified living environment, which has resulted in commercial pressures transcending the physical boundary of the home. Thus, instead of achieving the goal of attaining a work-life balance through running a home business, the changing business environment has led to the increasing commodification of the home and living spaces.

These findings have a number of practical implications. First, migrant entrepreneurs rent houses from local residents and thus raise the economic income of locals. Moreover, the guesthouses they operate become tourist attractions, which promote the tourism development of the destination. Migrant entrepreneurs also make important economic contributions to the local community (Bosworth & Farrell, 2011; Liu & Cheung, 2016), as they tend to have a higher education level and stronger economic capabilities than the local people. Nonetheless, their living status is not as good as the local residents and tourists imagine. Thus, local governments should attach greater importance to migrant entrepreneurs' business operations and living status, and set up a platform for lifestyle entrepreneurs to share information and provide training in business operations and management.

Second, although the tourism lifestyle industry is still small, it provides a rich context in which to explore social mobility. Future studies should explore the work and life role shifts and role conflicts of tourism lifestyle entrepreneurs in other cultural contexts to ascertain the cultural influences on the role transition dynamics (Ashforth et al., 2000; Hsieh & Lin, 2010). Moreover, it would be valuable to collect longitudinal data to investigate the dynamic nature of lifestyle entrepreneurs' role transitions. Finally, it would be interesting to conduct an in-depth investigation of the role shifting of different categories of lifestyle entrepreneurs, such as those of different gender and in different life-cycle stages, to determine the relationship between role transition and work and life satisfaction.

Disclosure statement

No potential conflict of interest was reported by the authors.

Funding

This work was supported by the National Natural Science Foundation of China [41701156].

ORCID

Honggang Xu http://orcid.org/0000-0003-1354-6654

References

Agarwal, S., & Lenka, U. (2015). Study on work-life balance of women entrepreneurs: Review and research agenda. *Industrial and Commercial Training*, *47*(7), 356–362.

Akintayo, D. I. (2010). Work-family role conflict and organizational commitment among industrial workers in Nigeria. *International Journal of Psychology & Counselling*, *2*(1), 1–8.

Allen, T. D., Herst, D. E., Bruck, C. S., & Sutton, M. (2000). Consequences associated with work-to-family conflict:

A review and agenda for future research. *Journal of Occupational Health Psychology*, *5*(2), 278.

Allen, V. L., & Van de Vliert, E. (1984). *Role transitions: Explorations and explanations*. New York, NY: Plenum Press.

Ashforth, B. E., Kreiner, G. E., & Fugate, M. (2000). All in a day's work: Boundaries and micro role transitions. *Academy of Management Review*, *25*(3), 472–491.

Ateljevic, I., & Doorne, S. (2000). 'Staying within the fence': Lifestyle entrepreneurship in tourism. *Journal of Sustainable Tourism*, *8*(5), 378–392.

Benson, M., & O'Reilly, K. (2009). Migration and the search for a better way of life: A critical exploration of lifestyle migration. *The Sociological Review*, *57*(4), 608–625.

Bosworth, G., & Farrell, H. (2011). Tourism entrepreneurs in Northumberland. *Annals of Tourism Research*, *38*(4), 1474–1494.

Braun, V., & Clarke, V. (2006). Using thematic analysis in psychology. *Qualitative Research in Psychology*, *3*(2), 77–101.

Bredvold, R., & Skålén, P. (2016). Lifestyle entrepreneurs and their identity construction: A study of the tourism industry. *Tourism Management*, *56*, 96–105.

Burke, P. J. (1991). Identity processes and social stress. *American Sociological Review*, *56*(6), 836–849.

Burr, W. R. (1972). Role transitions: A reformulation of theory. *Journal of Marriage & Family*, *34*(3), 407–416.

Butler, R. (2004). The tourism area life cycle in the twenty-first century. *A Companion to Tourism*, 159–170.

Carlsen, J., Morrison, A., & Weber, P. (2008). Lifestyle oriented small tourism firms. *Tourism Recreation Research*, *33*(3), 255–263.

Chen, J., & Chen, N. (2017). Beyond the everyday? Rethinking place meanings in tourism. *Tourism Geographies*, *19*(1), 9-26.

Clark, S. C. (2000). Work/family border theory: A new theory of work/family balance. *Human Relations*, *53*(6), 747–770.

Cohen, S. A., Duncan, T., & Thulemark, M. (2015). Lifestyle mobilities: The crossroads of travel, leisure and migration. *Mobilities*, *10*(1), 155–172.

Coverman, S. (1989). Role overload, role conflict, and stress: Addressing consequences of multiple role demands. *Social Forces*, *67*(4), 965–982.

Deakins, D., & Freel, M. (2006). *Entrepreneurship and small firms* (3rd ed.). London, UK: McGraw-Hill Education.

Desrochers, S., Hilton, J. M., & Larwood, L. (2005). Preliminary validation of the work-family integration-blurring scale. *Journal of Family Issues*, *26*(4), 442–466.

Duchscher, J. E. B. (2009). Transition shock: The initial stage of role adaptation for newly graduated registered nurses. *Journal of Advanced Nursing*, *65*(5), 1103–1113.

Gibson, H., & Yiannakis, A. (2002). Tourist roles: Needs and the lifecourse. *Annals of Tourism Research*, *29*(2), 358–383.

Goffman, E. (1959). *The presentation of self in everyday life*. Garden City, NY: Anchor.

Guest, D. E. (2002). Perspectives on the study of work-life balance. *Social Science Information*, *41*(2), 255–279.

Holmes, S., & Zimmer, I. (1994). The nature of the small firm: Understanding the motivations of growth and non-growth oriented owners. *Australian Journal of Management*, *19*(1), 97–120.

Hsieh, Y. C. (2010). Bed-and-breakfast innkeepers in the United States: When the boundary between work and personal life is blurred. *Journal of Human Resources in Hospitality & Tourism*, *9*(2), 200–217.

Hsieh, Y. C., & Lin, Y. H. (2010). Bed and breakfast operators' work and personal life balance: A cross-cultural comparison. *International Journal of Hospitality Management*, *29*(4), 576–581.

Jones, A. P., & Butler, M. C. (1980). A role transition approach to the stresses of organizationally induced family role disruption. *Journal of Marriage & Family*, *42*(2), 367–376.

Kahn, R. L., Wolfe, D. M., Quinn, R. P., Snoek, J. D., & Rosenthal, R. A. (1964). *Organizational stress: Studies in role conflict and ambiguity*. New York, NY: Wiley.

Katz, D., & Kahn, R. L. (1978). *The social psychology of organizations* (Vol. 2). New York, NY: Wiley.

Keeney, J., Boyd, E. M., Sinha, R., Westring, A. F., & Ryan, A. M. (2013). From "work–Family" to "work–Life": Broadening our conceptualization and measurement. *Journal of Vocational Behavior*, *82*(3), 221–237.

Kopelman, R. E., Greenhaus, J. H., & Connolly, T. F. (1983). A model of work, family, and interrole conflict: A construct validation study. *Organizational Behavior and Human Performance*, *32*(2), 198–215.

Kossek, E. E., Noe, R. A., & Demarr, B. J. (1999). Work-family role synthesis: Individual and organizational determinants. *International Journal of Conflict Management*, *10*(2), 102–129.

Li, Y., Miao, L., Zhao, X. Y., & Lehto, X. R. (2013). When family rooms become guest lounges: Work-family balance of B&B innkeepers. *International Journal of Hospitality Management*, *34*(1), 138–149.

Lin, J. H., Wong, J. Y., & Ho, C. H. (2015). The role of work-to-leisure conflict in promoting frontline employees' leisure satisfaction: Examining the job demand-control-support model. *International Journal of Contemporary Hospitality Management*, *27*(7), 1539–1555.

Lin, Y. S., Huang, W. S., Yang, C. T., & Chiang, M. J. (2014). Work–Leisure conflict and its associations with well-being: The roles of social support, leisure participation and job burnout. *Tourism Management*, *45*, 244–252.

Liu, S., & Cheung, L. T. O. (2016). Sense of place and tourism business development. *Tourism Geographies*, *18*(2), 1–20.

Lu, Y. L. (2008). Does Hukou also work? The household registration system with social stratification and mobility. *Social Sciences in China*, *1*, 149–162.

Lynch, P. A. (2005). The commercial home enterprise and host: A United Kingdom perspective. *International Journal of Hospitality Management*, *24*(4), 533–553.

Ma, S. Y. (2013). *From consumption to production: Study of the lifestyle tourism entrepreneurial migrants in dali historical town* (Unpublished Master's thesis). Sun Yat-sen University, Guangzhou, China.

Marchant, B. (2010, June). *The case of lifestyle entrepreneurs in Ireland; An examination of surf tourism entrepreneurs in Bundoran and Lahinch*. Paper presented at the Tourism and hospitality research in Ireland conference. Clare.

Marcketti, S. B., Niehm, L. S., & Fuloria, R. (2006). An exploratory study of lifestyle entrepreneurship and its relationship to life quality. *Family and Consumer Sciences Research Journal*, *34*(3), 241–259.

Matthews, R. A., Barnes-Farrell, J. L., & Bulger, C. A. (2010). Advancing measurement of work and family domain boundary characteristics. *Journal of Vocational Behavior*, *77*(3), 447–460.

Matthews, R. A., & Winkel, D. E. (2014). A longitudinal examination of role overload and work–Family conflict: The mediating role of interdomain transitions. *Journal of Organizational Behavior*, *35*(1), 72–91.

McIntosh, A. J., Lynch, P., & Sweeney, M. (2011). "My home is my castle": Defiance of the commercial homestay host in tourism. *Journal of Travel Research*, *49*(1), 509–519.

Meleis, A. I., Sawyer, L. M., Im, E. O., Hilfinger Messias, D. K., & Schumacher, K. (2000). Experiencing transitions: An emerging middle-range theory. *Advances in Nursing Science*, *23*(1), 12.

Michel, J. S., Mitchelson, J. K., Kotrba, L. M., LeBreton, J. M., & Baltes, B. B. (2009). A comparative test of work-family conflict models and critical examination of work-family linkages. *Journal of Vocational Behavior*, *74*(2), 199–218.

O'Reilly, K., & Benson, M. (2009). Lifestyle migration: Escaping to the good life? In M. Benson & K. O'Reilly (Eds.), *Lifestyle migrations: Expectations, aspirations and experiences* (pp. 1–13). Aldershot: Ashgate.

Petty, N. J., Thomson, O. P., & Stew, G. (2012). Ready for a paradigm shift? part 1: Introducing the philosophy of qualitative research. *Manual Therapy*, *17*(4), 267–274.

Polkinghorne, D. E. (2007). Validity issues in narrative research. *Qualitative Inquiry*, *13*(4), 471–486.

Rehman, S., & Azam Roomi, M. (2012). Gender and work-life balance: A phenomenological study of women entrepreneurs in Pakistan. *Journal of Small Business and Enterprise Development*, *19*(2), 209–228.

Saxena, G. (2015). Imagined relational capital: An analytical tool in considering small tourism firms' sociality. *Tourism Management*, *49*, 109–118.

Schmidt, S., Roesler, U., Kusserow, T., & Rau, R. (2014). Uncertainty in the workplace: Examining role ambiguity and role conflict, and their link to depression – A meta-analysis. *European Journal of Work and Organizational Psychology*, *23*(1), 91–106.

Shaw, G., & Williams, A. M. (2004). From lifestyle consumption to lifestyle production: Changing patterns of tourism entrepreneurship. In R. Thomas, (ed.) *Small firms in tourism: international perspective* (pp. 99–113).Oxford: Elsevier.

Shelton, L. M. (2006). Female entrepreneurs, work–Family conflict, and venture performance: New insights into the work–Family interface. *Journal of Small Business Management*, *44* (2), 285–297.

Stryker, S., & Macke, A. S. (1978). Status inconsistency and role conflict. *Annual Review of Sociology*, *4*(1), 57–90.

Sun, X. X., & Xu, H. G. (2017). Lifestyle tourism entrepreneurs' mobility motivations: A case study on Dali and Lijiang, China. *Tourism Management Perspectives*, *24*, 64–71.

Sweeney, M., Docherty-Hughes, J., & Lynch, P. (2018). Lifestyling entrepreneurs' sociological expressionism. *Annals of Tourism Research*, *69*, 90–100.

Turner, R. H. (1990). Role change. *Annual Review of Sociology*, *16*(1), 87–110.

Walker, E., Wang, C., & Redmond, J. (2008). Women and work-life balance: Is home-based business ownership the solution? *Equal Opportunities International*, *27*(3), 258–275.

Wang, S., Hung, K., & Bao, J. G. (2015). Is lifestyle tourism business in the age of commercialization just a dream? Challenges and remedies. *Journal of China Tourism Research*, *11*(1), 19–34.

White, N. R., & White, P. B. (2004). Travel as transition: Identity and place. *Annals of Tourism Research*, *31*(1), 200–218.

Williams, A. M., & Hall, C. M. (2000). Tourism and migration: New relationships between production and consumption. *Tourism Geographies*, *2*(1), 5–27.

Williams, A. M., Shaw, G., & Greenwood, J. (1989). From tourist to tourism entrepreneur, from consumption to production: Evidence from Cornwall, England. *Environment and Planning A*, *21*(12), 1639–1653.

Winkel, D. E., & Clayton, R. W. (2010). Transitioning between work and family roles as a function of boundary flexibility and role salience. *Journal of Vocational Behavior*, *76*(2), 336–343.

Xu, H. G., & Ma, S. Y. (2012). Regional environment of destination and the entrepreneurship of small tourism businesses: A case study of Dali and Lijiang of Yunnan Province. *Asia Pacific Journal of Tourism Research*, *19*(2), 144–161.

Xu, H. G., & Wu, Y. F. (2016). Lifestyle mobility in China: Context, perspective and prospects. *Mobilities*, *11*(4), 509–520.

Zhang, J., Xu, H., & Xing, W. (2017). The host–Guest interactions in ethnic tourism, Lijiang, China. *Current Issues in Tourism*, 20(7), 724–739.

Dining at luxury restaurants when traveling abroad: incorporating destination attitude into a luxury consumption value model

Norman Peng, Annie Chen and Kuang-Peng Hung

ABSTRACT

As luxury products have become increasingly accessible to middle-class consumers, many gastronomic tourists might be interested in dining at luxury restaurants when traveling. The purpose of this study is to explore gastronomic tourists' consumption at luxury restaurants by incorporating "destination attitude" into a luxury consumption value model. Three hundred ninety participants from Taiwan completed questionnaires, which revealed that perceived functional value, perceived hedonic value, and perceived symbolic/expressive value can significantly influence diners' restaurant satisfaction, which in turn can affect their destination attitude and behavioral intentions toward luxury restaurants.

Introduction

The luxury product industry has grown significantly since the 1980s (D'Arpizio, Levato, Kamel, & de Montgolfier, 2017). From the 1980s to 2006, sales in the global luxury goods industry increased more than fourfold – from US $70 billion to $264 billion (all monetary figures in this paper are in USD) – more than four times the value of the same industry in 1997 (D'Arpizio et al., 2017). The World Tourism Organization (2012) noted that gastronomic tourism constitutes an important segment within the tourism industry, as food has been an important medium to understand culture and/or society (Chang, Kivela, & Mak, 2010). Within the luxury industry, dining at luxury restaurants is a growing trend (Chen, Peng, & Hung, 2015; Jang & Namkung, 2009; Wu & Liang, 2009). Additionally, consumers are increasingly willing to try luxury restaurants when participating in tourism activities (Chipkin, 2016; Enskog, 2014). A luxury restaurant is defined as a full-service restaurant whose environment (services, décor, and atmosphere) and products (food and beverages) are carefully prepared and presented as unique, superior in quality, and conspicuous (Chen et al., 2015).

Gaps in the literature

Consumers' additional willingness to try luxury restaurants when participating in tourism activities brings new business opportunities and challenges for luxury product marketers, restaurateurs, and tourism operators. More importantly, the trend to dine at luxury restaurants when traveling provides researchers with an opportunity to narrow the gaps in the luxury product consumption literature and tourism studies. First, although identifying the factors that contribute to consumers' intentions to purchase luxury products is of interest to researchers, previous luxury consumption research has focused more on goods than on services (Chen & Peng, 2014; Han, Nunes, & Dreze, 2010; Hung et al., 2011; Shukla & Purani, 2012; Wiedmann, Hennigs, & Siebels, 2009; Yang & Mattila, 2016). Existing theories on luxury consumption might need to be adjusted when applied to service-based products because of the nature of services, such as their intangibility and variability.

Second, researchers have suggested that the appeal of luxury brands and products is rooted in their ability to signify luxury value both to users and to others (Gurzki & Woisetschläger, 2017). Scholars have collected evidence to suggest that there is a significant relationship between perceived luxury value and intention to consume luxury goods (Berthon, Pitt, Parent, & Berthon, 2009; Vigneron & Johnson, 2004; Wiedmann et al., 2009; Yang & Mattila, 2016). However, findings on the influence of luxury value have been inconsistent when applied to luxury services. For example, some researchers found that symbolic value has a significant impact on luxury hotel consumers, while others suggested that it has an statistically insignificant effect on luxury restaurant diners (Chen & Peng, 2014; Yang & Mattila, 2016). In addition, some findings demonstrated that experiential value is an important factor in relation to diners' restaurant satisfaction, while others revealed that experiential/hedonic value has an insignificant influence on consumers of luxury accessories (Shukla & Purani, 2012; Wu & Liang, 2009). This inconsistency undermines the reliability of perceived luxury value when examining luxury product consumption.

Third, destination attitude's influence on tourists' luxury service product purchase decisions is another area within the luxury products consumption literature that shows potential for further research. Mainly, additional research can be conducted to explore how tourists' attitudes toward a destination can be affected by their satisfaction with an activity such as dining at a luxury restaurant. Moreover, the findings in the luxury goods consumption literature have demonstrated that attitudes toward consumption settings play an important role in users' experiences (Moore, Doherty, & Doyle, 2010; Zhan & He, 2012). Nevertheless, the current literature on the consumption of luxury products has yet to fully explore the antecedents of consumers' attitudes toward consumption settings when the product in question is a service. Examining tourists' consumption of luxury restaurants considering its influence on their attitude toward the destination may add new insight to the tourism studies and luxury product consumption literature.

Fourth, tourists spend approximately 40% of their tourism budget on food (Horng, Liu, Chou, & Tsai, 2012). Although there is no study that has focused on gastronomic tourists' spending breakdown in particular, it is likely that they spend more than 40% of their travel budget on food and drinks because, by definition, gastronomic tourists are tourists who focus on the consumption of food and drinks and the enjoyment of good food and drinks (Horng & Tsai, 2010). Nevertheless, no known study has explored how spending on food and drinks can moderate gastronomic tourists' perception of their luxury restaurant dining experiences. This study addresses tourists' consumption of food and beverages in a more general sense. Such tourists differ from culinary tourists. The latter tend to pay more attention to cooking and to the style of food preparation in isolation (Horng & Tsai, 2010). Research on the influence of tourists' spending on food and drinks may be of particular interest to academic communities when the product offering represents an intangible emotional experience of high monetary value. In addition, it also provides insight to practitioners on how to be more effective when targeting and serving gastronomic tourists.

Aims, objectives, and intended contributions

The aim of this research is to contribute to the gastronomic tourism literature and luxury product consumption studies through testing a framework that examines gastronomic tourists' behavioral intentions toward luxury restaurants by incorporating a "destination attitude" variable into a consumption value model. The objectives of this study are as follows. First, this research plans to examine the influence of luxury restaurant perceived value (i.e. functional value, financial value, hedonic value, and symbolic/expressive value) on tourists' restaurant satisfaction. Second, this research plans to examine the relationships between luxury restaurant satisfaction, destination attitude, and behavioral intentions toward a restaurant. Third, this research aims to examine the ability of restaurant spending to moderate the influence of perceived luxury value on tourists' restaurant satisfaction. Fourth, this study intends to further explore the question of whether existing theories on luxury consumption may need to be adjusted when applied to service-based products because of their differences from physical goods. Finally, this study advances a number of implications for the consideration of hospitality and luxury service marketing practitioners.

Literature review – consumption value theory, luxury value in a service-based context, and destination attitude

To support this investigation, we adapt the consumption value theory. Consumption value has been the focus of marketers' attention since the late 1980s, when it was deemed a key determinant of consumers' purchase decisions (Sánchez, Callarisa, Rodrı´Guez, & Moliner, 2006). In particular, consumption value theory considers the benefits that consumers received from a product, such as social benefits, and the sacrifices they made, such as financial sacrifices. Furthermore, researchers have arrived at the consensus that consumption value has multiple dimensions and that both the utilitarian and non-utilitarian values of products should be considered (Sweeney & Soutar, 2001). For instance, Sheth, Newman, and Gross (1991) proposed that consumption value consists of social value, emotional value, functional value, epistemic value, and conditional value. Alternatively, Sweeney and Soutar (2001) suggest that consumption value includes functional dimensions (such as economic and quality), social dimensions, and emotional dimensions.

The consumption value model originated in marketing and has gradually been applied in other fields, including luxury product consumption behavior (Berthon et al., 2009; Hung et al., 2011; Vigneron & Johnson, 2004). When compared to non-luxury products, luxury products tend to have premium quality, recognizable style, strong reputation, and high hedonic value, in addition to being more expensive (Berthon et al., 2009). Thus, in those studies that examine luxury products' consumption value, researchers have suggested these products' particular characteristics must be taken into account (Vigneron & Johnson, 2004). For this reason, apart from the perceived functional and financial value associated with luxury products, researchers have stressed the importance of these products' ability to bring joy to their users (i.e. perceived hedonic value) and to symbolize wealth and status (i.e. perceived symbolic value) (e.g. Han et al., 2010; Wiedmann et al., 2009).

Although the previous literature has investigated luxury products' consumption value, only scant luxury product consumption research has focused on services. Of the scholarship that has examined luxury service's perceived value, Chen and Peng (2014), Wu and Liang (2009), and Yang and Mattila (2016) are the most relevant to the current research. Wu and Liang (2009) focus on the context of luxury restaurant consumption and examine service performance, the dining environment, and additional influences on perceived experiential value and satisfaction. Their results show that customers' perceived experiential value of luxury restaurants can affect their satisfaction toward luxury restaurants. Additionally, Wu and Liang (2009) provide an operational definition of luxury restaurants, which includes four-star and five-star hotels' full-service restaurants because they are subject to stringent inspection, high customer expectations, and competition from nearby restaurants.

Chen and Peng (2014) provide a glimpse into tourists' luxury hotel consumption by adapting the luxury value dimensions (i.e. functional value, experiential value, and symbolic value) developed by Berthon et al. (2009). Based on 368 Chinese tourists' responses, these authors showed that luxury hotels' perceived luxury value can affect tourists' attitudes, which, in turn, can affect their staying behavior. Moreover, symbolic value and experiential value can directly influence their staying behavior, but functional value does not. Chen and Peng (2014) suggested this might be the case because it is difficult for consumers to evaluate the functional benefits of luxury hotels prior to their stay.

When examining luxury services, such as Michelin-starred restaurants, Yang and Mattila (2016) proposed that consumers' luxury value has four dimensions: functional value, financial value, hedonic value, and symbolic/expressive value. Yang and Mattila showed that functional value, financial value, and hedonic value can positively and directly affect consumers' purchase intentions. Notably, they proposed that symbolic/expressive value will not affect purchase intentions. When consumers are seeking symbolic/expressive value, these authors argued that luxury goods are more suitable than luxury services because tangible goods are more visible.

The present study adopts the luxury value dimensions of Yang and Mattila (2016), as their study focuses on luxury consumption and is more comprehensive than those of Wu and Liang (2009) and Chen and Peng (2014). Wu and Liang (2009) only examined experiential value, whereas Chen and Peng (2014) did not consider financial value, which arguably is a key component of consumption value (Sheth et al., 1991; Sweeney & Soutar, 2001). Previous findings generated from luxury restaurant diners have been inconsistent (e.g. Wu & Liang, 2009; Yang & Mattila, 2016); therefore, this present study examines consumers' behavioral intentions toward luxury restaurants in order to provide additional insight into how perceived luxury value affects luxury service product consumers. Furthermore, although this study focuses on luxury restaurants, the context of this research is different from that of previous studies, as we investigate gastronomic tourists who dine at luxury restaurants when traveling aboard for tourism purposes. Previous studies mainly focused on consumers who reside in the same city or in nearby regions to where the restaurant is located (e.g. Chen et al., 2015; Jang & Namkung, 2009).

In this study, the perceived functional value of luxury restaurants measures the extent to which their products are desirable in terms of quality and performance (Yang & Mattila, 2016). Second, the hedonic value of luxury restaurants refers to their ability to arouse emotions of indulgence and pleasure in consumers (Yang & Mattila, 2016). Third, the symbolic/expressive value of luxury restaurants indicates their ability to express and relay information about customers' wealth and status to others and to the customers themselves (Hung et al., 2011). The financial value of a luxury restaurant relates to consumers' perceptions of the cost-benefit relationship involved in dining there (Wiedmann et al., 2009; Yang & Mattila, 2016).

Finally, this study considers the role of destination attitude in affecting tourists' behavioral intentions

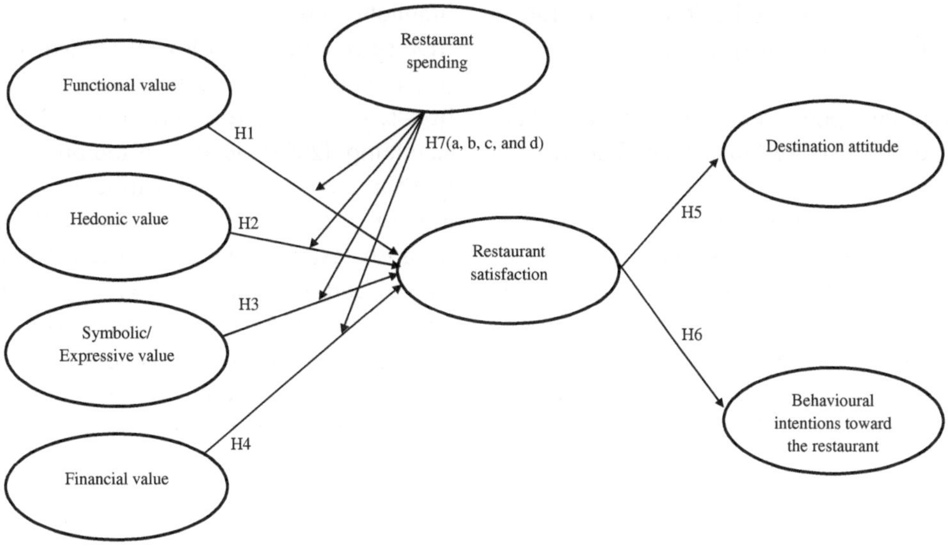

Figure 1. Main research framework.

toward luxury restaurants. In this study, destination attitude is defined as a tourist's overall affective appraisal of the city in which a luxury restaurant is located as a tourism destination (Ajzen, 1991). This factor could affect gastronomic tourists' future behavioral intentions, such as intentions to revisit, because tourism is generally considered to be a highly planned activity (Sirakaya & Woodside, 2005); therefore, it is less likely that tourists will want to revisit a luxury restaurant that is located in a city they do not like. From a theoretical perspective, studies have demonstrated that food and drink can contribute to tourists' perceptions of a destination (Hillel, Belhassen, & Shani, 2013; Kivela & Crotts, 2006; Lin, 2014; Sánchez-Cañizares & López-Guzmán, 2012). Although some studies have been conducted to examine how food and drink can affect tourists' perceptions of a destination, no known study has examined the relationships between gastronomic tourists' luxury restaurant satisfaction, their attitude toward the city in which the restaurant is located as a tourism destination, and their intentions to revisit and recommend a luxury restaurant.

Research framework and hypotheses

Based on a review of studies examining perceived luxury value and destination attitude, this paper adopts the model framework developed by Williams and Soutar (2009). This framework shows that perceived value can affect tourists' satisfaction, which, in turn, can affect their behavioral intentions. In terms of perceived luxury value, this research considers the influence of functional value, financial value, hedonic value, and symbolic/expressive value. Furthermore, the current study incorporates "destination attitude" into the proposed model to investigate its influence on tourists' restaurant satisfaction. The following section provides details regarding this study's proposed framework and hypotheses (Figure 1).

Research hypothesis – the influence of perceived luxury value

The first hypothesis examined by this research involves the influence of a luxury restaurant's perceived functional value on gastronomic tourists' restaurant satisfaction. Satisfaction was defined as gastronomic tourists' affective overall appraisal of the luxury restaurants visited (Dagger & David, 2012). In the consumer research literature, hospitality and tourism products' perceived functional value has consistently been found to have a significant impact on consumer satisfaction (Ha & Jang, 2010; Lee, Lee, & Choi, 2011; Williams & Soutar, 2009). Consumers will be satisfied if the luxury products they purchased are of superior quality (Berthon et al., 2009). In Kwun and Oh (2006) and Yang and Mattila (2016), the restaurant's functional value included the service it provides, its interior design, and the food and beverages served. Additionally, Jang and Namkung (2009) revealed that service quality, product quality, and atmosphere are essential components of a luxury restaurant's appeal. Diners will be pleased if luxury restaurants have a high functional value. Based on the works of Jang and Namkung (2009) and Kwun and Oh (2006), this research proposes that the satisfaction that gastronomic tourists derive from the luxury restaurants that they visit will be greater when the restaurant is perceived as having a high functional value, such as having tasty dishes.

H1: Luxury restaurants' perceived functional value has a positive effect on tourists' restaurant satisfaction.

The second hypothesis of this study focuses on the influence of perceived hedonic value on tourist's restaurant satisfaction. Berthon et al. (2009) and Vigneron and Johnson (2004) have argued that consumers purchase luxury goods because of the enjoyment that these products bring them. Luxury goods' ability to provide a sense of pleasure and self-indulgence has been suggested as a key reason why consumers purchase them (Berthon et al., 2009; Hung et al., 2011; Vigneron & Johnson, 2004). Hedonic value also plays an important role in luxury restaurant consumption (Chen et al., 2015; Jang & Namkung, 2009; Lee & Hwang, 2011; Wu & Liang, 2009). Luxury restaurant consumption behavior extends beyond fulfilling basic needs, as these experiences can come to becoming an art form that gratifies multiple human senses (Chang et al., 2010; Hillel et al., 2013; Kivela & Crotts, 2006; Lee & Hwang, 2011). In their research, Sirakaya-Turk, Ekinci, and Martin (2015) theorized and confirmed that hedonic value can influence tourists' shopping satisfaction. Wu and Liang (2009) showed that luxury restaurants can increase diners' levels of satisfaction by gratifying their senses. This study hypothesizes that gastronomic tourists' restaurant satisfaction can be positively affected by luxury restaurants' perceived hedonic value, such as by enhancing diners' sense of enjoyment.

H2: Luxury restaurants' perceived hedonic value has a positive effect on tourists' restaurant satisfaction.

The third hypothesis tested by this research concerns the influence of the symbolic/expressive value on gastronomic tourists' restaurant satisfaction. Luxury products can signal users' wealth and status to other users, among others (Yang & Mattila, 2016). In addition, consumers can use these goods and services to express their values and beliefs to others and to themselves (Gurzki & Woisetschläger, 2017; Hung et al., 2011; O'Cass, 2004). Shukla and Purani (2012) and Chen and Peng (2014) found that this variable is significant with regard to consumers' overall evaluation of luxury goods and services. Hung et al. (2011) suggested that luxury products' symbolic value is one of the main reasons that consumers purchase these products.

In the case of luxury services, Chen and Peng (2014) found that lodgers perceive a hotel as good/pleasant if they believe that this hospitality service will potentially highlight their status and wealth to others and to themselves. To attract luxury service consumers, luxury hotels need to be more conspicuous than non-luxury hotels. According to Chen et al. (2015) and Jang and Namkung (2009), luxury restaurants employ multiple methods to embed their symbolic value (e.g. luxurious interior environments and dishes with rare ingredients) because restaurateurs believe that these methods can lead to better evaluations from customers. The research participants in these studies showed the significance of luxury restaurants' conspicuousness. Based on the works of the aforementioned authors, the present study proposes that luxury restaurants' perceived symbolic value will have a significant impact on customers' satisfaction with these restaurants. In other words, diners will perceive their dining experiences as having exceeded their expectations if they believe that these restaurants are able to relay their status and wealth to others and/or to themselves.

H3: Luxury restaurants' perceived symbolic/expressive value has a positive effect on tourists' restaurant satisfaction.

The fourth hypothesis investigates the influence of luxury restaurants' perceived financial value on gastronomic tourists' restaurant satisfaction. Consumption value theory suggests that consumers will weigh the benefits derived from a product against the sacrifices made to obtain such product, such as financial sacrifices. Hennigs et al. (2012), Shukla and Purani (2012) and Wiedmann et al. (2009) demonstrated the importance of perceived financial value in the context of luxury product consumption. Some luxury brand managers will justify their products' high prices by suggesting luxury products will last longer, perform better, and/or be a better investment than non-luxury products (Wiedmann et al., 2009). In other words, although consumers usually make more financial sacrifices when purchasing luxury products than when purchasing non-luxury products, these additional sacrifices can be justified if these products are perceived as having more benefits than their non-luxury counterparts.

The relevance of perceived financial value has also received some attention in the context of luxury dining. Harrington, Ottenbacher, and Kendall (2011) suggested that appropriate price and value for money are two important attributes consumers consider when selecting fine dining restaurants. As diners attempt to maximize benefits and minimize costs when dining at luxury restaurants, they are concerned with issues such as whether the cost is justified. However, the value of luxury restaurants is more than food that satisfies diners' basic needs; it includes superior atmospherics and product/service quality as well (Chen et al., 2015). Based on the work by Chen et al. (2015) and Harrington et al. (2011), the value of luxury restaurants from a financial perspective might have a significant impact on tourists' satisfaction with a restaurant. This study proposes that gastronomic tourists will be pleased with their visit if they believe dining at a luxury restaurant offers them more benefits than the financial scarifies that they made.

H4: Luxury restaurants' perceived financial value has a positive effect on tourists' restaurant satisfaction.

Research hypotheses – the influences of satisfaction

The fifth relationship examined in this study is the effect of restaurant satisfaction on destination attitude. Gastronomic tourists are known to be interested in engaging in tourism activities while enjoying good food and drink (Horng & Tsai, 2010), and a consumer's attitude toward the focal place and destination might be conditioned by the activities that he/she participated in, such as dining at luxury restaurants (Okumus, Okumus, & McKercher, 2007; Sánchez-Cañizares & López-Guzmán, 2012; Tsai & Wang, 2017).

Research has suggested that food and beverages, particularly when treated as an experience, in general may contribute to tourists' perceptions of a destination (Björk & Kauppinen-Räisänen, 2014, 2016; Hillel et al., 2013; Kivela & Crotts, 2006; Lin, 2014; Sánchez-Cañizares & López-Guzmán, 2012). For example, Du Rand, Heath, and Alberts (2003) theorized and confirmed that food can be a key and/or supporting attraction for potential tourists to South Africa. Little empirical research has examined how tourists' satisfaction with a restaurant may affect their attitude toward the city in which it is located as a tourism destination. Oliver (1981) has argued that consumers' satisfaction can be an antecedent to their attitude, and Hsu, Cai, and Li (2010) argued that satisfaction is an input to attitude. This study extends the current hospitality and luxury brand consumption literature by proposing that gastronomic tourists will perceive the cities in which the restaurants are located as exciting, arousing, pleasant, and/or relaxing if they are satisfied with the luxury restaurants that they visit. We hypothesize that gastronomic tourists' restaurant satisfaction will have a positive influence on their attitude toward the cities in which the restaurants are located.

H5: Restaurant satisfaction has a positive effect on tourists' attitude toward a destination.

The sixth hypothesis addressed in this study involves the influence of restaurant satisfaction on gastronomic tourists' behavioral intentions toward luxury restaurants. In the present research, "behavioral intentions" is defined as a deeply held commitment to consistently revisit and recommend a preferred luxury restaurant (Oliver, 1999). Theories that study consumer behavior, such as social exchange theory and self-congruence theory, have consistently found consumer satisfaction to be a key indicator to consumers' future behavioral intentions (Jin, 2012; Nam, Ekinci, & Whyatt, 2011; Shiau & Luo, 2012; Wu, 2013). In the tourism literature, scholars have found that tourists' restaurant satisfaction with the tourism products they consume, such as local food, affects their behavioral intentions to, for instance, revisit the same establishment and/or to recommend it to others (Quintal & Polczynski, 2010; Um, Chon, & Ro, 2006; Wan & Chan, 2013; Williams & Soutar, 2009). Based on the findings from Um et al. (2006) and Wan and Chan (2013), this research proposes that gastronomic tourists will be more likely to revisit a luxury restaurant and recommend it to others if they are highly satisfied with their restaurant visit.

H6: Restaurant satisfaction has a positive effect on tourists' behavioral intentions toward the restaurant, as measured by intentions to revisit and to recommend a luxury restaurant to others.

The seventh hypothesis tests the ability of gastronomic tourists' restaurant spending to moderate the influence of the perceived luxury value on their restaurant satisfaction. A review of the existing tourist behavior literature reveals no known study that has examined whether spending can moderate a luxury restaurant's perceived value and diners' satisfaction. However, there is circumstantial evidence to support this hypothesis (Alegre & Juaneda, 2006; Han & Hyun, 2015; Matzler, Füller, & Faullant, 2007; Ryu & Han, 2010; Wong & Dioko, 2013).

From a theoretical point of view, it is reasonable to expect spending level to moderate perceived value's influence on customer satisfaction. Matzler et al. (2007) propose that spending levels (i.e. high versus low) will moderate the attribute performance-satisfaction-loyalty chain in the tourism context. Their findings show that ski resorts' product attributes have greater influence on customer satisfaction for tourists who spend more on this experience than for tourists who spend less (Matzler et al., 2007). Alegre and Juaneda (2006) also confirm that spending/expenditure is a useful factor to differentiate first-time visitors to a destination from repeaters.

Wong and Dioko (2013) extend the disconfirmation paradigm literature by suggesting casino customers' expectations might be a moderator of the relationships between perceived performance, perceived value, and customer satisfaction. Their result confirms that perceived value's influence on satisfaction can be positively moderated by customers' expectations. It aligns with Zeithaml, Berry, and Parasuraman's (1993) suggestion that customers with high expectations have a narrow tolerance zone and are easier to delight when the performance and value of services meet their desired service level. According to the disconfirmation paradigm literature, customers might experience positive or negative disconfirmation after comparing their expectations with a product's perceived value (Oliver, 1980). Some researchers suggest satisfaction is the outcome of a comparison between expectations and the perceived performance of a product or service (Wong & Dioko, 2013).

Other than confirming expectation's moderating effect, Wong and Dioko point out that the content and influence of consumer expectations are often subject to contextual variations. In the case of luxury product consumption and hospitality services, spending is often a key indicator of consumers' expectations and subsequent behavioral intentions (Alegre & Juaneda, 2006; Berthon et al., 2009; Han & Hyun, 2015; Ryu & Han, 2010; Wiedmann et al., 2009); therefore, this research explores spending level's ability to moderate the influence of luxury restaurants' perceived luxury value on gastronomic tourist's satisfaction. Considering spending level as an important indicator of luxury product consumers' expectations, studies have confirmed that expectations can moderate the relationship between perceived value and satisfaction, and existing literature on the disconfirmation paradigm (Matzler et al., 2007; Wong & Dioko, 2013; Zeithaml et al., 1993), it is reasonable to assume that the influence of the perceived luxury value on gastronomic tourists' restaurant satisfaction will be more intense for tourists who spend more money on this experience than for those who spend less. This study thus proposes hypothesis H7.

H7: Perceived luxury value (a. perceived functional value, b. perceived hedonic value, c. perceived symbolic/expressive value, and d. perceived financial value) has a stronger relationship with restaurant satisfaction for tourists who spend more in luxury restaurants than for tourists who spend less in luxury restaurants.

Method

Sampling and data collection methods

To examine the proposed framework, this study focused on Taiwanese consumers who dined at luxury restaurants when they traveled abroad independently (as opposed to joining a group package tour) as tourists. Taiwanese consumers were suitable for exploring this study's research questions for three reasons. First, Taiwanese consumers have shown that they are enthusiastic about luxury restaurants (Chen et al., 2015; Wu & Liang, 2009). Second, Taiwanese consumers are interested in trying out local food when they are tourists (Chang, 2009; Chang, Kivela, & Mak, 2011). Third, Taiwanese consumers enjoy traveling for tourism purposes. During 2014 and 2015, one out of four Taiwanese individuals made at least one international trip, and the average trip duration for Taiwanese tourists was more than 8 nights (Chang, 2009; Tourism Bureau, 2015). Considering these developments, Taiwanese tourists will have multiple opportunities to dine in foreign countries. Some will be motivated to patronize luxury restaurants when participating in tourism activities.

Trained interviewers distributed and collected the questionnaires. A non-probability sampling method was employed in this research. A purposive sampling technique was used to collect the data. This sampling method was suitable because there is no complete directory of gastronomic tourists. In other words, the members of this specific group were impossible to identify; therefore, randomly selecting participants was not feasible (Bhattacherjee, 2012). Using an interception technique, the interviewers selected individuals who arrived at Taiwan's Taoyuan International Airport and at Taipei Songshan Airport (Wong & Yeh, 2009). This strategy was applied to increase the chances of meeting individuals who had dined at luxury restaurants when traveling abroad for tourism purposes.

The respondents for this research were Taiwanese tourists who had dined at luxury restaurants during their previous international trip for leisure purposes independently (as opposed to joining a group package tour). In addition, potential participants had to be over the age of 18, pay for the meal themselves, and have an interest in gastronomic tourism activities. In other words, respondents had to have engaged in food-related activities because they found such activities to be somewhat interesting and different – not simply because they were hungry (Tsai & Wang, 2017). These were screening questions that the interviewer asked potential participants at the start of the interview.

Table 1. Characteristics of the participants (N = 390).

Variables	Demographic traits	%
Gender	Male	59.2
	Female	40.8
Respondent's age	Between 18–30 years old	35.1
	Between 31–40 years old	31
	Between 41–50 years old	19.2
	Between 51–60 years old	10.3
	61 and above	4.4
Education	High school degree	8
	College degree	25.4
	University	44.4
	Postgraduate degree or above	22.2

The customers were asked to participate in the survey on a voluntary basis. To ensure that participants understood the context of this research, a description of luxury restaurants adapted from Wu and Liang (2009) and Yang and Mattila (2016) and examples of luxury restaurants were presented to the participants at the beginning of the interview. Finally, the participants were asked to give a brief description of a luxury restaurant that they had visited as tourists, including the name of the restaurant and its signature dishes.

After twelve weeks of data collection, 390 usable questionnaires were collected. The number of usable surveys in this study were in line with that of other studies of luxury/upscale restaurants that used an on-site data collection method (Chen et al., 2015; Jang & Namkung, 2009; Ryu, Lee, & Kim, 2012). Each diner spent an average of approximately $99 per meal per person, ranging between $60 and $200 (excluding service charges and tips). This pricing is higher than that defined by Chen et al. (2015) for luxury restaurants ($67). The demographic breakdown of the sample set can be found in Table 1.

Questionnaire design

The participants completed a 22-question survey that included three items for functional value (Yang & Mattila, 2016), three items for symbolic/expressive value (Yang & Mattila, 2016), three items for hedonic value (Yang & Mattila, 2016), three items for financial value (Yang & Mattila, 2016), three items for restaurant satisfaction (Taplin, 2013), three items for behavioral intentions toward luxury restaurants (Jang & Namkung, 2009), and four items for destination attitude (Carlson, Rosenberger III, & Rahman, 2015). Because Jang and Namkung (2009), Yang and Mattila (2016), Taplin (2013), and Carlson et al. (2015) used a seven-point Likert-type scale, a seven-point Likert-type scale was used in this study when designing the items. All of the variables in the model were measured using multiple items and were found to be reliable, with Cronbach's alpha values ranging from 0.76 to 0.89. The items for each variable are presented in Table 2.

Data analysis and results

Model measurement

IBM SPSS AMOS 23 was used to analyze the data. A two-step approach to structural equation modeling

Table 2. Measurement items.

Variable/Adopted from	Measurement items
Functional value (FuV)/Yang and Mattila (2016)	FuV1: Luxury Restaurant X is aesthetically appealing.
	FuV2: Luxury Restaurant X's dishes are sophisticated.
	FuV3: The service provided in Luxury Restaurant X is attentive.
Symbolic/Expressive value (SEV)/Yang and Mattila (2016)	SEV1: Dining at Luxury Restaurant X is considered a symbol of social status.
	SEV2: Dining at Luxury Restaurant X helps me to express myself.
	SEV3: Dining at Luxury Restaurant X helps me communicate my self-identity.
Hedonic value (HV)/Yang and Mattila (2016)	HV1: I dine at Luxury Restaurant X for the pure enjoyment of it.
	HV2: Dining at Luxury Restaurant X is self-indulging.
	HV3: Dining at Luxury Restaurant X gives me a lot of pleasure.
Financial value (FiV)/Yang and Mattila (2016)	FiV1: It is worth the economy investment to dine at Luxury Restaurant X.
	FiV2: Dining at Luxury Restaurant X is worth its high price.
	FiV3: Luxury Restaurant X offer value for money.
Destination attitude (DA)/Carlson et al. (2015)	City X[2] was–
	DA1: Gloomy/Exciting[3]
	DA2: Unpleasant/Pleasant[3]
	DA3: Sleepy/Arousing[3]
	DA4: Distressing/Relaxing[3]
Restaurant satisfaction (S)/Taplin (2013)	RS1: I was satisfied with my visit to Luxury Restaurant X.
	RS2: My expectations for this visit were exceeded.
	RS3: I am pleased with this visit to Luxury Restaurant X.
Behavioral intentions towards the restaurant (BI)/Jang and Namkung (2009)	BI1: I would like to come back to Luxury Restaurant X in the future.
	BI2: I plan to revisit Luxury Restaurant X in the future.
	BI3: I would recommend Luxury Restaurant X to my friends or others.

1. In the survey, "Luxury Restaurant X" is the restaurant respondent visited.
2. In the survey, "City X" is the city in which Luxury Restaurant X is located.
3. A series of seven semantic differential scales were used.

(SEM) was used, as recommended by Anderson and Gerbing (1988). All the factor loadings on the intended latent variables were found to be significant and greater than 0.7 (Fornell & Larcker, 1981). The squared multiple correlations supported the reliability of the measurement items. Convergent validity was examined in terms of factor loadings and average variance extracted (AVE). AVE is the average variance shared between a construct and its measurement (Fornell & Larcker, 1981), and the AVE values ranged from 0.53% to 0.74% (Table 3). Therefore, convergent validity was established (Fornell & Larcker, 1981).

Discriminant validity was evaluated by comparing the AVE of each individual construct with the shared variances between the individual construct and all the other constructs. Discriminant validity was upheld, as the AVE value for each construct was greater than the squared correlation between constructs (Table 3).

Table 3. Correlations and descriptive statistics.

	Mean	SD	CrA	CR	AVE	FuV	HV	SEV	FiV	DA	RS	BI
FuV	5.45	1.11	.89	.89	.74	**.86**						
HV	5.04	1.17	.89	.89	.74	.49	**.86**					
SEV	4.79	1.07	.76	.77	.53	.42	.64	**.73**				
FiV	4.77	1.19	.89	.89	.74	.44	.51	.49	**.86**			
DA	5.18	1.32	.85	.90	.70	.37	.41	.38	.41	**.84**		
RS	4.85	1.17	.89	.86	.68	.49	.46	.37	.39	.51	**.82**	
BI	5.46	1.13	.77	.78	.55	.16	.16	.21	.20	.30	.18	**.74**

‾Bold numbers on the diagonal parentheses are square root of each construct's AVE value
‾CrA = Cronach's Alphas; CR = Composite reliability; AVE = Average variance extracted
-FuV = functional value; HV = hedonic value; SEV = symbolic/expressive value; FiV = Financial value DA = destination attitude; RS = Restaurant satisfaction; BI = behavioral intentions toward the restaurant

Common method variance was checked using Harman's single-factor test. An unrotated factor analysis generated all the factors with eigenvalues greater than one. Notably, the first factor accounted for 44.8% of the variance (<50%), indicating that common method bias was unlikely to be a concern (Podsakoff, MacKenzie, Jeong-Yeon, & Podsakoff, 2003).

Structural model

After the overall measurement model was found to be acceptable, the structural model was examined. The model fit was good (χ^2 = 546.1; *df* = 197; χ^2/*df* = 2.772; RMSEA = 0.067; CFI = 0.934; NFI = 0.901), and the results obtained from examining the proposed hypotheses are presented in Figure 2. Hypothesis H1 was supported (t = 4.53; β = 0.24; $p < 0.001$) because the perceived functional value of luxury restaurants has a positive impact on gastronomic tourists' restaurant satisfaction. Hypothesis H2 was also supported (t = 2.33; β = 0.18; $p < 0.05$), as perceived hedonic value has a positive impact on tourists' restaurant satisfaction. Hypothesis H3 was supported (t = 4.60; β = .22; $p < 0.001$), suggesting that perceived symbolic/expressive value significantly affects tourists' restaurant satisfaction. Hypothesis H4 posited that perceived financial value would positively influence tourists' restaurant satisfaction. The result gathered from the statistical analysis did not support this hypothesis (t = –.03; β = –.29; $p > 0.1$). Hypothesis H5 was supported (t = 8.04; β = 0.54; $p < 0.001$). Tourists' restaurant satisfaction had a positive impact on their attitudes

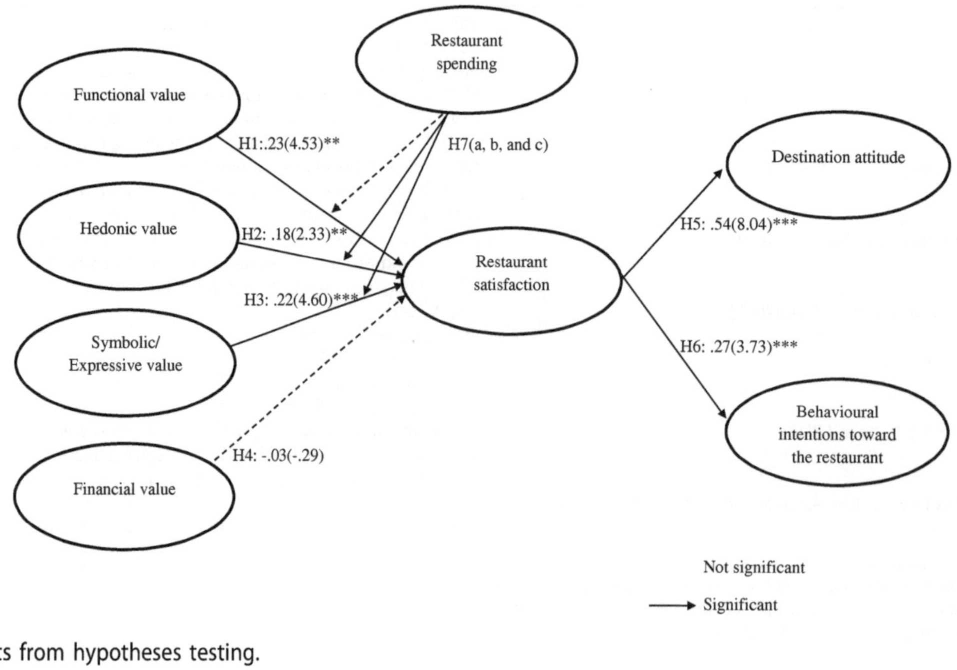

Figure 2. Results from hypotheses testing.

Table 4. Two group path model estimate (H7).

Path estimated	Customer spending in restaurant: Low spending group	Customer spending in restaurant: High spending group	($\Delta\chi^2$, $\Delta df = 1$)	Moderating effect
H7a: Perceived functional value→ Restaurant satisfaction	.19*	.22***	0.78	Not significant
H7b: Perceived hedonic value→ Restaurant satisfaction	.10	.31***	2.8	Significant
H7c: Perceived symbolic/expressive value→ Restaurant satisfaction	.10	.32***	7.65	Significant
H7d: Perceived financial value→ Restaurant satisfaction	No further test conducted as perceived financial value's direct influence on restaurant satisfaction (H4) was not supported			

- *$p < .05$. **$p < .01$. ***$p < .001$.

toward the city in which the luxury restaurant is located as a tourism destination. Hypothesis H6 was supported (t = 3.73; β = 0.27; p < 0.001), as tourists' restaurant satisfaction had a significant impact on their behavioral intentions toward the restaurant. The R^2 for this research's observed variable (destination attitude and behavioral intentions) are 0.47 and 0.87, respectively.

The moderating effect of gastronomic tourists' restaurant spending (H7)

To examine the moderating effect of gastronomic tourists' restaurant spending, a multigroup analysis was performed. The respondents were divided into two groups, a high spending group and a low spending group, using means split before the analysis. The mean of the participants' spending was 5.18 on a seven-point Likert-type scale; therefore, those with spending means greater than 5.18 were categorized in the high restaurant spending group (N = 226) and those with means below 5.18 were categorized in the low restaurant spending group (N = 164).

To test the differential effect between the high spending group and the low spending group, the chi-square difference between the constrained and unconstrained models was assessed by the difference in degrees of freedom (Anderson & Gerbing, 1988). In the unconstrained model, all path coefficients in each group were freely estimated. In the constrained model, the path coefficients for the relationships between perceived functional value and satisfaction, perceived symbolic/expressive value and satisfaction, and perceived hedonic value and satisfaction were set equally across the two groups. The chi-square difference (Δchi-square = 30.9, Δdf = 5) between the constrained model (chi-square = 769.7, df = 293) and the unconstrained model (chi-square = 738.8, df = 288) was significant ($p = 0.024$). The results demonstrated that the high spending group and the low spending group were significantly different on the model level.

To identify where the statistical differences appear, coefficients for each path were compared between the two groups (Table 4). Gastronomic tourists' restaurant spending did not moderate the relationship between perceived functional value and satisfaction. The difference in the coefficients in H7a (between perceived functional value and satisfaction) was not significant (Δchi-square = 0.78, Δdf = 1, $p > 0.05$) between the constrained model (chi-square = 739.58, df = 289) and the unconstrained model (chi-square = 738.8, df = 288). Statistical tests revealed that the moderating role of gastronomic tourists' restaurant spending existed on two hypothesized paths. First, the difference in the coefficients in H7b (between perceived hedonic value and satisfaction) was significant at (Δchi-square = 2.8, Δdf = 1, $p < 0.05$) between the constrained model (chi-square = 741.6, df = 289) and the unconstrained model (chi-square = 738.8, df = 288). Second, the difference in the coefficients for H7c (between perceived symbolic/expressive value and satisfaction) was significant (Δchi-square = 7.65, Δdf = 1, $p < 0.001$) between the constrained model (chi-square = 746.45, df = 289) and the unconstrained model (chi-square = 738.8, df = 288). Because perceived financial value had an insignificant impact on tourists' restaurant satisfaction, H7d was not examined. Based on the above analysis, H7 is partially supported.

Discussion and practical implications

Discussion

The findings of this study are generally consistent with those in the literature on consumption value, i.e. that consumers are more likely to be satisfied if they perceive that the products that they have purchased have value (e.g. utility). However, this study also contributes to the luxury product consumption and gastronomic tourism literature in five ways. First, the results that were obtained in this research provide insight on the question of whether existing theories on luxury consumption may need to be adjusted when applied to service-based products because of their differences from physical goods. By applying the luxury consumption value model to a tourist luxury restaurant consumption scenario, it could be argued that there are both overlapping areas (i.e. the influence of perceived hedonic value, perceived symbolic/expressive value, and perceived functional value) and differences (i.e. the unexpected insignificant effects of perceived financial value); therefore, when applying research frameworks to examine luxury goods consumption behavior to luxury service consumption behavior, researchers may have to keep this issue in mind.

Second, this study provides additional insights into the relationship between gastronomic tourists' restaurant satisfaction, behavioral intentions toward restaurants, and destination attitude. This study's result demonstrates that satisfaction toward a luxury restaurant has a significant effect on gastronomic tourists' intentions to revisit and to recommend a restaurant. In other words, having meals that exceeded gastronomic tourists' expectations was sufficient to stimulate their intentions to revisit and to recommend a restaurant. This result is similar to those of previous consumption studies that suggest that satisfaction has a direct impact on behavioral intentions (e.g. Quintal & Polczynski, 2010; Um et al., 2006; Wan & Chan, 2013).

Not only does tourists' restaurant satisfaction have a significant influence on behavioral intentions toward the restaurant, but also it can affect tourists' attitudes toward the cities in which the restaurants are located as tourism destinations. Previous studies have suggested that food and beverages in general may contribute to tourists' perceptions of a destination (Hillel et al., 2013; Kivela & Crotts, 2006; Lin, 2014; Sánchez-Cañizares & López-Guzmán, 2012); however, few studies have examined the relationship between restaurant satisfaction and destination attitude. This research's finding shows a consumer's attitude toward the destination might be conditioned by the activities that he/she participated in, such as dining at luxury restaurants. Destination attitude's R^2 (0.47) can be interpreted as a tourist's dining satisfaction having a reasonable amount of power to explain his/her attitude toward a destination.

Third, this study contributes to the debate regarding symbolic value's influence on the consumers of luxury services. Unlike studies on luxury goods, which have received more attention from researchers and for which the findings have been more consistent (Han et al., 2010; Hung et al., 2011; Shukla & Purani, 2012; Wiedmann et al., 2009), there are fewer studies on the consumption of luxury services, and the findings from those extant studies have been less consistent (e.g. Chen & Peng, 2014; Jang & Namkung, 2009; Yang & Mattila, 2016). This study's finding aligns with the results of Chen and Peng (2014): symbolic/expressive value has a significant influence on gastronomic tourists' restaurant satisfaction. Although luxury services might be less visible than luxury goods (Yang & Mattilia, 2016), luxury restaurants can still signal diners' wealth and status to others and to the diners themselves. In addition, wealth and status do not always have to be signaled by luxury fashion items or luxury cars that can be carried or driven around. They can also be expressed through restaurant choices, even when the restaurant experience occurs abroad.

Fourth, the luxury restaurant's perceived functional value and hedonic value both positively influenced gastronomic tourists' restaurant satisfaction. These results align with the previous luxury restaurant research on diners who are not tourists (e.g. Jang & Namkung, 2009; Kwun & Oh, 2006; Wu & Liang, 2009). Because this research focuses on restaurants' perceived value, this study's approach can be argued to bypass each hospitality product's unique facility requirements, providing additional implications for theory and practice compared to previous studies (Jang & Namkung, 2009; Kwun & Oh, 2006). Notably, this research does not support the claim that a luxury restaurant's perceived financial value has a significant impact on gastronomic tourists' restaurant satisfaction. Petrick (2005) revealed in a study of passengers on a cruise that those passengers who choose high-end cruise lines rather than value-oriented cruise lines were less sensitive about price. It is possible that gastronomic tourists who choose to dine at luxury restaurants when going abroad are less sensitive to the luxury restaurant's price or whether it is a good economic investment given that they are aware it is an experience that lasts only a few hours and is intangible. However, this interpretation will require additional research.

Fifth, perhaps one of this study's most significant findings relates to the moderating effect of tourists' spending in luxury restaurants on the relationship between perceived luxury value and restaurant satisfaction. The inability of tourists' spending to moderate the influence of the perceived functional value on restaurant satisfaction demonstrates the importance of a product's functional value. Quality food and drinks are important to gastronomic tourists' satisfaction with luxury restaurants, regardless of their spending levels.

The relationship between perceived hedonic value and restaurant satisfaction and that between perceived symbolic/expressive value and restaurant satisfaction are significantly different between gastronomic tourists who spend more in a restaurant and those who spend less. In other words, the relationship between perceived luxury value and satisfaction intensifies as gastronomic tourists increase their spending in luxury restaurants. This finding confirms that tourists' spending in luxury restaurants serves as an effective moderator of these relationships. Gastronomic tourists who spent greater amounts of money in luxury restaurants pay closer attention to whether this service will signify their wealth and status to others and to themselves. Moreover, they are more responsive to the joy and happiness this experience can bring them. Additional research is required to pinpoint the cause of this difference, but one possible explanation is that these consumers might have higher levels of expectation and/or vanity than their counterparts. Following this discussion of this study's contributions to the literature, the next section will present the practical implications of this research.

Practical implications

How to build a holistic experience that connects food and tourism and how to market and promote gastronomic tourism are two important questions for tourism bureaus and restaurants that want to attract more gastronomic tourists (World Tourism Organization, 2012). In addition to offering theoretical contributions, this study has several managerial implications for practitioners to consider.

First and foremost, luxury restaurants can expect gastronomic tourists to recommend and revisit their establishment if they were satisfied with their dining experiences. Therefore, luxury restaurant managers who want gastronomic tourists to revisit and recommend their restaurants may have to evaluate tourists' expectations and find out whether they were pleased with their visit. A self-assessment based on management experience may be a good starting point. Alternatively, managers could ask diners to fill out a survey after they have finished their meals.

Second, gastronomic tourism constitutes an important segment within the tourism industry (World Tourism Organization, 2012), and the restaurant industry in general has made significant contributions to the economy (National Restaurant Association, 2016). Restaurant managers can make a contribution to tourists' attitudes toward the cities in which the restaurants in question are located. For those cities/regions that tourists currently do not find to be relaxing, exciting, or pleasant but want to attract more future gastronomic tourists, this study recommends that restaurant professionals and tourism bureaus in the region collaborate with one another by launching a rebranding campaign. Promoting and advertising in the countries that send the most tourists may be one option. Alternatively, targeting countries with tourists who are known for enjoying ethnic cuisine may also be useful.

Last, as consumers are increasingly more willing to try luxury restaurants when participating in tourism activities (Chipkin, 2016; Enskog, 2014), luxury restaurants will need to have some knowledge on how they are similar to and different from diners who are not tourists. As is the case with luxury restaurant customers who are not tourists, enjoying excellent food and drink made with premium ingredients is essential to gastronomic tourists' satisfaction. In addition, being able to please tourists through superior services and esthetics can also contribute to gastronomic tourists' satisfaction and subsequent behavioral intentions. Moreover, similar to those concerned with appealing to consumers of luxury products such as cars, fashion accessories, and hotels, restaurateurs must consider whether their products and services can signal diners' wealth and status because this study finds that perceived symbolic/expressive value has a significant effect on tourists' satisfaction.

Restaurants will have an idea of gastronomic tourists' level of spending once they order. Regardless of their level of spending, high-quality food is key to customer satisfaction, but consumers who spend more are different from those who spend less in terms of how they perceive a restaurant's symbolic/expressive value and its hedonic value. At appropriate times, such as when introducing the menu and during the short breaks between dishes, service staff could consider incorporating a few words/gestures to encourage those customers who are likely prepared to spend beyond average in the restaurant to signify their social status and wealth. It might also be worthwhile to have the manager/owner and head chef, especially for restaurants with celebrity chefs, drop by and greet the customers personally. Managers could also consider providing some goodwill gestures, such as complementary dishes or desserts, to customers who are likely prepared to spend more than average at the restaurant. This treatment should intensify these customers' satisfaction, which in turn will contribute to their positive word-of-mouth and revisit intentions.

Although restaurant practitioners should not take advantage of gastronomic tourists because they may not be sensitive to luxury restaurants' financial value – such as whether the meal provides value for money – when serving diners who are traveling abroad, they can consider including additional delectable, high-quality dishes that have higher profit margins, such as common poultry dishes that may bring tourists great satisfaction.

Conclusions

Luxury services have received little attention from scholars who study luxury product consumption.

Furthermore, the influence of perceived luxury value in the literature has been inconsistent. To contribute to the literature and practice, "destination attitude" was incorporated into a luxury consumption value model to conceptualize gastronomic tourists' behavioral intentions toward restaurants. In terms of the factors contributing to gastronomic tourists' satisfaction, perceived functional value, perceived symbolic/expressive value, and perceived hedonic value were shown to have significant effects. Additionally, perceived financial value had no significant impact on gastronomic tourists' restaurant satisfaction. Furthermore, gastronomic tourists' satisfaction has a significant effect on their behavioral intentions toward a restaurant and destination attitude.

Despite its contributions, this study has several limitations. First, Bei and Chiao (2001) and Xia, Monroe, and Cox (2004) proposed that perceived price fairness can have a significant impact on diners' satisfaction and loyalty; nevertheless, this study did not support the contention that perceived financial value, which shares some similarity with perceived price fairness, has a significant impact on diners' satisfaction. Researchers may wish to further explore the reasons behind this outcome. Second, this study included gastronomic tourists who have engaged in food-related activities because they found such activities to be somewhat interesting and different. Through this approach, it is reasonable to expect that experiences with food and drink will be a major factor that contributes to our participants' attitudes toward the destinations that they visited. Nevertheless, destination attitude could be affected by other non-food-related factors as well. Future studies might want to consider comparing how gastronomic tourists and other types of tourists formulate their attitudes toward a destination.

Third, tourists formulate their perceptions of a destination through affective appraisal and cognitive evaluation. This study examined "destination attitude". Future studies might want to explore factors that contribute to tourists' attitudes toward a destination, such as perceived destination image. Finally, a luxury restaurant's aesthetic has been considered to be part of a luxury restaurant's experiential value (Wu & Liang, 2009) and part of its functional value (Yang & Mattila, 2016). It is possible that aesthetics is both part of a luxury restaurant's experience and part of its product. However, experiential value's influence was not examined in this research. Future studies could explore how experiential value and perceived luxury value might affect luxury restaurant customers similarly and differently.

Disclosure statement

No potential conflict of interest was reported by the authors.

References

Ajzen, I. (1991). The theory of planned behavior. *Organizational Behavior and Human Decision Processes, 50*, 179–211.

Alegre, J., & Juaneda, C. (2006). Destination loyalty. Consumers' economic behavior. *Annals of Tourism Research, 33*(3), 684–706.

Anderson, J. C., & Gerbing, D. W. (1988). Structural equation modelling in practice: A review and recommended two-step approach. *Psychological Bulletin, 103*(3), 411–423.

Bei, L. T., & Chiao, Y. C. (2001). An integrated model for the effects of perceived product, perceived service quality, and perceived price fairness on consumer satisfaction and loyalty. *Journal of Consumer Satisfaction, Dissatisfaction and Complaining Behavior, 14*, 125–140.

Berthon, P., Pitt, L., Parent, M., & Berthon, J.-P. (2009). Aesthetics and ephemerality: Observing and preserving the luxury brand. *California Management Review, 52*(1), 45–66.

Bhattacherjee, A. (2012). *Social science research: Principles, methods, and practices. Textbooks collection. Book 3.* Retrieved from http://scholarcommons.usf.edu/oa_textbooks/3

Björk, P., & Kauppinen-Räisänen, H. (2014). Culinary-gastronomic tourism- a search for local food experiences. *Nutrition & Food Science, 44*(4), 294–309.

Björk, P., & Kauppinen-Räisänen, H. (2016). Local food: A source for destination attraction. *International Journal of Contemporary Hospitality Management, 28*(1), 177–194.

Carlson, J., Rosenberger III, P. J., & Rahman, M. M. (2015). Cultivating group-oriented travel behaviour to major events: Assessing the importance of customer-perceived value, enduring event involvement and attitude towards the host destination. *Journal of Marketing Management, 31*(9–10), 1065–1089.

Chang, J.-C. (2009). Taiwanese tourists' perceptions of service quality on outbound guided package tours: A qualitative examination of the SERVQUAL dimensions. *Journal of Vacation Marketing, 15*(2), 165–178.

Chang, R. C. Y., Kivela, J., & Mak, A. H. N. (2010). Food preferences of Chinese tourists. *Annals of Tourism Research, 37*(4), 989–1011.

Chang, R. C. Y., Kivela, J., & Mak, A. H. N. (2011). Attributes that influence the evaluation of travel dining experience: When East meets West. *Tourism Management, 32*(2), 307–316.

Chen, A., & Peng, N. (2014). Examining Chinese consumers' luxury hotel staying behavior. *International Journal of Hospitality Management, 39*, 53–56.

Chen, A., Peng, N., & Hung, K.-P. (2015). The effects of luxury restaurant environments on diners' emotions and loyalty: Incorporating diner expectations into an extended Mehrabian-Russell model. *International Journal of Contemporary Hospitality Management, 27*(2), 236–260.

Chipkin, H. (2016). Travel professionals identify 15h luxury travel trends for 2016 and beyond. Retrieved May 8, 2018, from http://www.travelmarketreport.com/articles/Travel-

Professionals-Identify-15-Luxury-Travel-Trends-For-2016-And-Beyond.

D'Arpizio, C., Levato, F., Kamel, M.-A., & de Montgolfier, J. (2017). Luxury goods worldwide market study. Retrieved April 9, from http://www.bain.com/publications/articles/luxury-goods-worldwide-market-study-fall-winter-2017.aspx

Dagger, T. S., & David, M. E. (2012). Uncovering the real effect of switching costs on the satisfaction-loyalty association. *European Journal of Marketing*, *46*(3/4), 447–468.

Du Rand, G. E., Heath, E., & Alberts, N. (2003). The role of local and regional food in destination marketing- A South African situation analysis. *Journal of Travel & Tourism Marketing*, *14*(3–4), 97–112.

Enskog, D. (2014). Luxury tourism- The Top 7 trends. Retrieved May 8, 2018, from https://www.credit-suisse.com/us/en/articles/articles/news-and-expertise/2014/01/en/luxury-tourism-the-top-7-trends.html

Fornell, C., & Larcker, D. (1981). Evaluating structural equation models with unobservable variables and measurement errors. *Journal of Marketing Research*, *18*(1), 39–50.

Gurzki, H., & Woisetschläger, D. M. (2017). Mapping the luxury research landscape: A bibliometric citation analysis. *Journal of Business Research*, *77*, 147–166.

Ha, J., & Jang, S. (2010). Perceived values, satisfaction, and behavioral intentions: The role of familiarity in Korean restaurants. *International Journal of Hospitality Management*, *29*, 2–13.

Han, H., & Hyun, S. S. (2015). Customer retention in the medical tourism industry: Impact of quality, satisfaction, trust, and price reasonableness. *Tourism Management*, *46*, 20–29.

Han, Y. J., Nunes, J. C., & Dreze, X. (2010). Signaling status with luxury goods: The role of brand prominence. *Journal of Marketing*, *74*(4), 15–30.

Harrington, R. J., Ottenbacher, M. C., & Kendall, K. W. (2011). Fine-dining restaurant selection: Direct and moderating effects of customer attributes. *Journal of Foodservice Business Research*, *14*(3), 272–289.

Hennigs, N., Wiedmann, K.-P., Klarmann, C., Godey, B., Pederzoli, D., Neulinger, A., … Oh, H. (2012). What is the value of luxury? A cross-cultural consumer perspective. *Psychology & Marketing*, *29*(12), 1018–1034.

Hillel, D., Belhassen, Y., & Shani, A. (2013). What makes a gastronomic destination attractive? Evidence from the Israeli Negev. *Tourism Management*, *36*, 200–209.

Horng, J.-S., Liu, C.-H. S., Chou, H.-Y., & Tsai, C.-Y. (2012). Understanding the impact of culinary brand equity and destination familiarity on travel intentions. *Tourism Management*, *33*(4), 815–824.

Horng, J.-S., & Tsai, C.-T. (. S. (2010). Government websites for promoting East Asian culinary tourism: A cross-national analysis. *Tourism Management*, *31*, 74–85.

Hsu, C. H. C., Cai, L. A., & Li, M. (2010). Expectation, motivation, and attitude: A tourist behavioral model. *Journal of Travel Research*, *49*(3), 282–296.

Hung, K.-P., Chen, A. H., Peng, N., Hackley, C., Tiwsaku, A. R., & Chou, C.-L. (2011). Antecedents of luxury brand purchase intention. *Journal of Product & Brand Management*, *20*(6), 456–467.

Jang, S. (S), & Namkung, Y. (2009). Perceived quality, emotions, and behavioural intentions: Application of an extended Mehrabian-Russell model to restaurants. *Journal of Business Research*, *62*(4), 451–460.

Jin, S.-A.-A. (2012). The potential of social media for luxury brand management. *Marketing Intelligence & Planning*, *30*(7), 687–699.

Kivela, J., & Crotts, J. C. (2006). Tourism and gastronomy: Gastronomy's influence on how tourists' experience a destination. *Journal of Hospitality & Tourism Research*, *30*(3), 354–377.

Kwun, D. J.-W., & Oh, H. (2006). Past experience and self-image in fine dining intentions. *Journal of Foodservice Business Research*, *9*(4), 3–23.

Lee, J. H., & Hwang, J. (2011). Luxury marketing: The influences of psychological and demographic characteristics on attitudes toward luxury restaurants. *International Journal of Hospitality Management*, *30*(3), 658–669.

Lee, J.-S., Lee, C.-K., & Choi, Y. (2011). Examining the role of emotional and functional values in festival evaluation. *Journal of Travel Research*, *50*, 685–696.

Lin, C.-H. (2014). Effects of cuisine experience, psychological well-being, and self-health perception on the revisit intention of hot springs tourists. *Journal of Hospitality & Tourism Research*, *38*(2), 243–265.

Matzler, K., Füller, J., & Faullant, R. (2007). Customer satisfaction and loyalty to Alpine ski resorts: The moderating effect of lifestyle, spending and customers' skiing skills. *International Journal of Tourism Research*, *9*, 409–421.

Moore, C. M., Doherty, A. M., & Doyle, S. A. (2010). Flagship stores as a market entry method: The perspective of luxury fashion retailing. *European Journal of Marketing*, *44*(1/2), 139–161.

Nam, J., Ekinci, Y., & Whyatt, G. (2011). Brand equity, brand loyalty and consumer satisfaction. *Annals of Tourism Research*, *38*(3), 1009–1030.

National Restaurant Association. (2016). 2016 Restaurant Industry Forecast. Retrieved September 22, 2017 from http://www.restaurant.org/Home/old_files/2016-Restaurant-Industry-Forecast.

O'Cass, A. (2004). Fashion clothing consumption: Antecedents and consequences of fashion clothing involvement. *European Journal of Marketing*, *38*(7), 869–882.

Okumus, B., Okumus, F., & McKercher, B. (2007). Incorporating local and international cuisines in the marketing of tourism destinations: The cases of Hong Kong and Turkey. *Tourism Management*, *28*, 253–261.

Oliver, R. L. (1980). A cognitive model of the antecedents and consequences of satisfaction decisions. *Journal of Marketing Research*, *17*(4), 460–469.

Oliver, R. L. (1981). Measurement and evaluation of satisfaction processes in retail settings. *Journal of Retailing*, *57*(3), 25–48.

Oliver, R. L. (1999). Whence consumer loyalty? *Journal of Marketing*, *63*, 33–44.

Petrick, J. F. (2005). Segmenting cruise passengers with price sensitivity. *Tourism Management*, *26*, 753–762.

Podsakoff, P., MacKenzie, S. L., Jeong-Yeon, L., & Podsakoff, N. (2003). Common method biases in behavioral research: A critical review of the literature and recommended remedies. *Journal of Applied Psychology*, *88*, 879–903.

Quintal, V. A., & Polczynski, A. (2010). Factors influencing tourists' revisit intentions. *Asia Pacific Journal of Marketing and Logistics*, *22*(4), 554–578.

Ryu, K., & Han, H. (2010). Influence of the quality of food, service, and physical environment on customer satisfaction and behavioral intention in quick-casual restaurants: Moderating role of perceived price. *Journal of Hospitality & Tourism Research*, *34*(3), 310–329.

Ryu, K., Lee, H.-R., & Kim, W. (2012). The influence of the quality of the physical environment, food, and service on restaurant image, customer perceived value, customer satisfaction, and behavioral intentions. *International Journal of Contemporary Hospitality Management*, *24*(2), 200–223.

Sánchez, J., Callarisa, L., Rodrı´Guez, L. M., & Moliner, M. A. (2006). Perceived value of the purchase of a tourism product. *Tourism Management*, *27*(3), 394–409.

Sánchez-Cañizares, S. M., & López-Guzmán, T. (2012). Gastronomy as a tourism resource: Profile of the culinary tourist. *Current Issues in Tourism*, *15*(3), 229–245.

Sheth, J. N., Newman, B. I., & Gross, B. L. (1991). Why we buy what we buy: A theory of consumption values. *Journal of Business*, *22*(2), 159–170.

Shiau, W.-L., & Luo, M. M. (2012). Factors affecting online group buying intention and satisfaction: A social exchange theory perspective. *Computers in Human Behavior*, *28*(6), 2431–2444.

Shukla, P., & Purani, K. (2012). Comparing the importance of luxury value perceptions in cross-national contexts. *Journal of Business Research*, *65*, 1417–1424.

Sirakaya, E., & Woodside, A. (2005). Building and testing theories of decision making by travelers. *Tourism Management*, *26*(6), 815–832.

Sirakaya-Turk, E., Ekinci, Y., & Martin, D. (2015). The efficacy of shopping value in predicting destination loyalty. *Journal of Business Research*, *68*, 1878–1885.

Sweeney, J. C., & Soutar, G. N. (2001). Consumer perceived value: The development of a multiple item scale. *Journal of Retailing*, *77*(2), 203–220.

Taplin, R. H. (2013). The influence of competition on visitor satisfaction and loyalty. *Tourism Management*, *36*, 238–246.

Tourism Bureau. (2015). Republic of China 104- An investigation and report of citizen's tourism condition. Retrieved October 31, 2016, from http://admin.taiwan.net.tw/upload/statistic/20160810/5edad339-d16f-4933-982d-9c90c72f5739.pdf.

Tsai, C.-T. (S), & Wang, Y.-C. (2017). Experiential value in branding food tourism. *Journal of Destination Marketing & Management*, *6*, 56–65.

Um, S., Chon, K., & Ro, Y. (2006). Antecedents of revisit intention. *Annals of Tourism Research*, *33*(4), 1141–1158.

Vigneron, F., & Johnson, L. W. (2004). Measuring perceptions of brand luxury. *Journal of Brand Management*, *11*(6), 484–506.

Wan, Y. K. P., & Chan, S. H. J. (2013). Factors that affect the levels of tourists' satisfaction and loyalty towards food festivals: A case study of Macau. *International Journal of Tourism Research*, *15*(3), 226–240.

Wiedmann, K.-P., Hennigs, N., & Siebels, A. (2009). Value-based segmentation of luxury consumption behaviour. *Psychology and Marketing*, *26*(7), 625–651.

Williams, P., & Soutar, G. N. (2009). Value, satisfaction and behavioral intentions in an adventure tourism context. *Annals of Tourism Research*, *36*(3), 413–438.

Wong, I. A., & Dioko, L. A. D. (D). (2013). Understanding the mediated moderating role of customer expectations in the customer satisfaction model: The case of casinos. *Tourism Management*, *36*, 188–199.

Wong, J.-Y., & Yeh, C. (2009). Tourist hesitation in destination decision making. *Annals of Tourism Research*, *36*(1), 6–23.

World Tourism Organization. (2012). Global report on food tourism. Retrieved October 31, 2016 from http://cf.cdn.unwto.org/sites/all/files/pdf/food_tourism_ok.pdf.

Wu, C. H.-J., & Liang, R.-D. (2009). Effect of experiential value on customer satisfaction with service encounters in luxury-hotel restaurants. *International Journal of Hospitality Management*, *28*, 586–593.

Wu, I.-L. (2013). The antecedents of customer satisfaction and its link to complaint intentions in online shopping: An integration of justice, technology, and trust. *International Journal of Information Management*, *33*, 166–176.

Xia, L., Monroe, K. B., & Cox, J. L. (2004). The price is unfair! A conceptual framework of price fairness perceptions. *Journal of Marketing*, *68*(4), 1–15.

Yang, W., & Mattila, A. S. (2016). Why do we buy luxury experiences? *International Journal of Contemporary Hospitality Management*, *28*(9), 1848–1867.

Zeithaml, V. A., Berry, L. L., & Parasuraman, A. (1993). The nature and determinants of customer expectations of services. *Journal of the Academy of Marketing Science*, *21*(1), 1–12.

Zhan, L., & He, Y. (2012). Understanding luxury consumption in China: Consumer perceptions of best-known brands. *Journal of Business Research*, *65*(10), 1452–1460.

China watching: luxury consumption and its implications

Kam Hung, Hanqin Qiu Zhang, Basak Denizci Guillet and Liang Wang

ABSTRACT

This study seeks to understand luxury shopping destinations of Mainland Chinese residents. Using Hong Kong as an example, this study conducted five focus group discussions and identified brand and high prices as proxies of luxury in the Chinese mindset. Findings that motivations for luxury consumption are socially and personally oriented simultaneously indicate that the framework of the self-concept theory is appropriate. However, the development of a luxury shopping destination goes beyond fulfilling the respondents' social and personal desires for luxury. The halo effects of social movements, such as Occupy Central and anti-parallel trade protests, have immensely mitigated the attractiveness of Hong Kong as a luxury shopping destination.

Introduction

Shopping as a pivotal touristic activity has been well-recognized in extant literature (Jin, Moscardo, & Murphy, 2017). The equation "leisure + shopping = tourism product mix" proposed by Jansen-Verbeke (1990) vividly depicts the significance of shopping in shaping tourism experiences. The term "shopping tourism" was recently coined as a distinctive form of tourism for which shopping is the primary travel purpose (Timothy, 2005). In this category, shopping is the main activity for travel and determines the visited locations (i.e. destination choice) (Wong & Wan, 2013). For destinations, shopping functions as a major source of revenue, and destination marketers have incorporated shopping in promotional campaigns (Rabbiosi, 2015). The role of shopping in destination selection has also attracted considerable academic attention. Yeung, Wong, and Ko (2004) compared Hong Kong to Singapore in terms of its attraction for shoppers. Their results indicated that the status of Hong Kong as the shopping paradise of the Asia Pacific region is being seriously challenged by Singapore. Shopping values are also effective indicators of destination loyalty. Sirakaya-Turk, Ekinci, and Martin (2015) suggested that hedonic and utilitarian shopping values are strongly correlated to shopping satisfaction, thereby mediating the effects of shopping values on destination loyalty.

Various items ranging from functional goods (e.g. running shoes) to expensive merchandise (e.g. antique watches and jewelry) fall under the term "souvenir." Souvenir purchase has become a central topic among shopping tourism-related studies, such as those related to influencing factors (Albayrak, Caber, & Çömen, 2016) and items purchased from a unique tourism destination (e.g. North Korea) (Li & Ryan, 2018). This factor is unsurprising because souvenirs have sensory and symbolic values. Souvenirs remind tourists of their touristic experiences and are given to others to maintain social networks (Swanson, 2014). These studies treated shopping as a byproduct of visiting a destination instead of as a travel purpose. Shopping tourism has gradually become common in the current tourism market. Shopping for luxury goods is also an important factor that motivates people to travel. Park, Reisinger, and Noh (2010) determined that individuals who attach importance to shopping cared about luxury goods. Therefore, arguing that luxury shopping is an important stimulus factor is reasonable. Occasionally, this factor is the only reason people visit a destination rather than being a mere touristic activity of a refined visiting experience in a place away from home.

Extant literature has extensively covered the stimuli-based destination selection from various perspectives (e.g. Choi, Liu, & Kim, 2015; Mussalam & Tajeddini, 2016; Sirakaya-Turk et al., 2015). However, only few academic studies have specifically been devoted to analyzing the role of luxury shopping in destination selection (Correia, Kozak, & Kim, 2017, 2018; Hung, Guillet, & Zhang, 2018). The current study aims to fill this void by using Hong

Kong as an example to explore the perceptions of luxury shopping destinations among Mainland Chinese residents. The limited literature that analyzes luxury shopping destinations prompted the aim of this study, namely to explore first an understanding of luxury and the motivations for luxury consumption. Thereafter, the perceptions of a specific luxury shopping destination are examined to identify the influencing factors beyond the identified elements centering on luxury consumption. Through these two steps, this study determines the considerations involved in luxury shopping destination selection by integrating luxury consumption into a substantially dynamic and complex tourism context. In accordance with the explorative nature of the research questions, this study adopts a qualitative research technique of focus group discussions for data collection.

In order to address the research question, two reasons are provided in analyzing Mainland Chinese residents' perceptions of Hong Kong as a luxury shopping destination. First, Hong Kong has been the top short-haul destination for Mainland Chinese travelers for a long time because of its geographical proximity to the Mainland. For certain Chinese consumers, shopping is the sole purpose of their visit (Heung & Qu, 1998). Second, the attractiveness of Hong Kong as a popular shopping destination is being challenged by increasing competition from nearby countries, such as Japan and South Korea, or areas as far as Europe and the United States (US). The changing preference of the Chinese in luxury consumption for tangibles and intangibles, including travel, can be attributed to various factors. Insights into the underlying motivations for pursuing luxury consumption will clarify the marketing and management practices of the luxury goods industry and destinations to address the needs of this emerging market in more detail.

Third, Mainland Chinese travelers are well suited to a study on the connections between luxury shopping and destination choice because, in contrast to their counterparts in Japan who prefer shopping for luxury goods domestically, the Chinese tend to purchase luxury goods overseas as they spend three times more in other countries than they do locally (Bain & Company, 2015).*The Economist* (2015a) echoed this statement and observed that approximately two-thirds of luxury goods purchased by the Chinese are bought outside China. Hence, the Chinese market has a natural advantage when it comes to discussing luxury shopping destinations because Chinese consumers' purchase of luxury items is closely connected to temporary mobility. Such an initiative is complementary to the current studies on luxury consumption and destination image among Chinese consumers. A few of these studies integrate luxury shopping into the destination image formation process.

Previous studies have also emphasized the critical roles of the socio-economic environment in understanding the consumer behavior of luxury consumption (e.g. Vickers & Renand, 2003). That is, luxury is a concept that may have varied meanings based on the socio-economic environment (Christodoulides, Michaelidou, & Li, 2009; Vickers & Renand, 2003). Park et al. (2010) suggested that the global luxury goods market is primarily driven by the growth of economies and high net-worth individuals. As a result of its accumulated material wealth, China has become the new arena for practicing materialistic values and conspicuous consumption, which have long been adopted as Western ideals (Podoshen, Li, & Zhang, 2011). Although the emergence of luxury consumption in China principally originates from rapid economic affluence, this market has its own distinct characteristics because consumer behavior is significantly modeled by factors along the spectrum of culture and personal features (Jap, 2010).

Literature review

This section presents findings from extant literature on luxury consumption. In particular, studies that analyze luxury per se, self-theory-based motivations for luxury consumption, and the role of China as an increasingly dominant player in the luxury shopping market are critically reviewed.

Concept of luxury

Understanding the nature of luxury requires consideration of its definition. The *Oxford Latin Dictionary* (1969) defines luxury as "something that is not essential but provides pleasure and comfort," while (*The Free Dictionary,* 2015) describes it as "something that is desirable but expensive or hard to obtain or do." Both interpretations highlight the nature of the non-necessity and premium price inherent in luxury consumption.

An analysis of the literature reveals a lack of consensus on the definition of luxury, which can be partially attributed to the fact that "luxury" per se is a vague term (Beverland, 2004). Scholars interchangeably use several concepts, such as luxury, luxury consumption, luxury items, and luxury goods (Han, Nunes, & Drèze, 2010; Lee & Hwang, 2011; Wong & Ahuvia, 1998). Another reason involves the divergent disciplinary perspectives adopted by researchers, with only a few of them attempting to reconcile various interpretations (Eng & Bogaert, 2010; Li, Li, & Kambele, 2012; Zhan & He, 2012). For example, Nueno and Quelch (1998, p. 62) adopted the economic perspective of defining luxury items as "those whose ratio of functional utility to price is low while the ratio of intangible and situational utility to price is high." Such an economic stance is consistent with the four consumption categorizations proposed by Adam Smith (i.e. necessary goods to maintain life; basic goods for normal growth and prosperity; affluence goods that are unnecessary for normal growth and prosperity; and luxury goods with limited supply, difficult accessibility, and high price) and stresses rarity and high price (Berthon, Pitt, Parent, & Berthon, 2009; Zhang & Kim, 2013). Goody (2006, p. 341) suggested that luxury implies "refined enjoyment, of elegance, of things desirable but not essential." Such an idea is relative to "daily living needs" (Vickers & Renand, 2003, p. 461).

Contrary to the conceptualization from an economic perspective, luxury has been conceptualized by a few scholars through the lens of social psychology theory, with emphasis placed on the conspicuous and social values inherent in the term "luxury" (Vigneron & Johnson, 1999). This perspective views the ownership of luxury items as an indicator of social status, thereby demonstrating that luxury plays a critical role in reconstructing social stratification (Kapferer & Bastien, 2009). Zhan and He (2012) believed that luxury consumption is inextricably linked to psychological benefits. Accordingly, psychological traits are fundamental in determining country-based differences in luxury consumption. Studies that adopted social psychology theories also reiterated the importance of interactions among members of a certain social community (Childers & Rao, 1992; Megehee & Spake, 2012). These interactions which are not necessarily actual or face-to-face contacts often represent the conceived influences from an actual or imaginary entity (i.e., either a person or a group). Such a phenomenon has been described as the bandwagon effect by Kastanakis and Balabanis (2012) and was referenced by Li and Su (2007).

Unlike these sections where carriers of luxury mainly include tangible items, such as watches and jewelry, the tourism industry is a novel domain for discussing the concept of luxury because it offers tangible and heterogeneous services (Litvin, Goldsmith, & Pan,2008). In the seminal book *Theory of the Leisure Class*, Veblen defined the leisure class in terms of consumption practices of travel, sports, and the arts (cited in Dimanche & Samdahl, 1994). Luxury holidays go beyond physical goods and are featured in specialized holiday offerings (Atwal & Williams, 2009). The narratives on symbolic consumption and tourism can be traced back to the study of Smith (2012), who investigated the reasons why middle-class Americans visited prestigious resorts (Ekinci, Sirakaya-Turk, & Preciado, 2013). Since then, academic inquiries into luxury in the tourism context have continued. Recent research on luxury in the tourism context has become substantially diversified because scholars shifted their attention from tourism per se as a luxury activity to decision attributes of the meaning of luxury. Chen and Peng (2014) studied the behaviors of Chinese consumers during their luxury hotel stays and verified a model of luxury-value attitudes. Hwang and Han (2014) analyzed the strategies for maximizing and utilizing brand prestige in the luxury cruise industry. They concluded that eight attributes affected the prestige of a luxury cruise brand. However, these tourism studies have overlooked a pervasive element: can the search for luxury be a dominant factor that propels tourists to embark on their trips? This question can be translated into discussions on destination image in relation to the places for luxury experiences or products. The present study uses luxury shopping to exploratively address this question, resulting in the determination of the perception of luxury in the targeted market. This study also ascertains the image of destinations by precisely fulfilling the consumers' expectations.

Motivation for luxury consumption: self-concept theory

The above-mentioned two preceding perspectives are consistent with the views relating to the driving factors of luxury consumption. These perspectives have been presented as divergent in the literature (Eng & Bogaert, 2010; Truong & McColl, 2011). Two major views exist and are sorted in a general manner. Both views are concerned with the motivations for luxury consumption, namely socially and individually oriented motivations. These motivations can be traced back to the notion of self-concept (Kastanakis & Balabanis, 2012). Numerous studies have suggested that consumption carries symbolic meanings because the items that consumers own can be recognized as part of their extended selves (Elliott & Wattanasuwan, 1998). This statement suggests the rationality of understanding consumer behavior within the self-concept framework.

Self-concept is referred to as the "totality of the individual's thoughts and feelings having references to himself as an object" (Rosenberg as cited in Sirgy, 1982), and it has been adopted in consumer behavior in various contexts, such as manufacturing (Heath & Scott, 1998) and service (Alden, He, & Chen, 2010). Kastanakis and Balabanis (2012) analyzed the impact of self-concept orientations upon bandwagon luxury consumption. Three mediating personality traits are included in this type of consumption: consumers' status-seeking predispositions, susceptibility to normative influence, and need for uniqueness. The results of the aforementioned study suggested that an interdependent self-concept underlies bandwagon luxury consumption. The interdependent self sees the self as an outer public self, which is exemplified in Asian cultures (e.g. Chinese and Japanese cultures). By contrast, the independent self perceives the self as an inner private self, which is dominant in Western cultures (Markus & Kitayama, 1991; Wong & Ahuvia, 1998). Self-concept theories have also been applied to understand luxury consumption types, particularly using the importance of price and behavior perceptions. For instance, one study revealed five categories of luxury consumers: Veblenian, snob, bandwagon, hedonist, and perfectionist (Vigneron & Johnson, 1999). Husic and Cicic (2009) further analyzed the underlying drivers for these five types of luxury consumer, namely brand image and quality, fashion, store atmosphere, and patron status.

Tourism scholars have also adopted the self-concept theory in recognizing the reasons for selecting certain products or destinations over others (Beerli, Meneses, & Gil, 2007; Sirgy & Su, 2000). This idea implies the long-held belief that the activity of tourism per se is symbolic, although such a halo effect has been fading in recent decades with the flourishing of low-cost carriers and global economic development. Chon (1992) and Hung and Petrick (2011) discussed self congruity in the tourism context and revealed that the higher the agreement between self-concept and destination image, the greater the satisfaction and travel intention of tourists. Todd (2001) claimed that self-concept can be an alternative segmentation base, thereby possibly enhancing the understanding of how tourists perceive themselves and their behaviors. Beerli et al. (2007) clarified the congruity between tourists' self-concept and destination image, thus positively correlating self-concept with the tendency of tourists to visit a certain place.

However, these studies overlooked the critical point that tourists are typically drawn in by a type of attraction, even though destination is intricately a comprehensive concept that consists of various attractions. This argument is implied in the scope of relevant research, such as shopping (Moscardo, 2004), culinary tourism (Ab Karim & Chi, 2010), and branding cultural festivals (Esu & Arrey, 2009) as destination attractions. Investigations of destination image are generally cascaded down to a single stimulus. Hence, resolutions based on the treatment of destination are less case-relevant than others. Moreover, these investigations did not specify the elements that constitute the interdependent and independent selves. This gap stops pertinent studies from having practical significance because culture contributes to global self-concept inconsistency (Spencer-Rodgers, Boucher, Mori, Wang, & Peng, 2009). Therefore, substantial academic attention should be devoted to comparing the perceptions of an image stimulus, such as luxury shopping in this research, with such a stimulus-based destination image.

Methodology

Data collection process: focus group discussion

A focus group interview was used in this study because it allows for an open-ended discussion on a certain topic among a group of respondents. Although several scholars (Calder, 1977) regard this technique as preliminary because of concerns about its subjectivity, previous studies have also demonstrated its usefulness in obtaining substantive information on the motivations and perceptions of participants (e.g. Huang & Hsu, 2005). Hence, a focus group discussion was appropriate for achieving the proposed research objectives given the exploratory nature of the current study. This work followed Hennink, Hutter, and Bailey (2011) in organizing the focus group interview process, which involves developing the discussion guide, selecting respondents, and conducting the discussion.

After a thorough review of the relevant literature, open-ended questions pertaining to issues such as motivations in luxury shopping and perceptions of Hong Kong as a luxury shopping destination were used throughout the discussion. Most of the questions were psychologically and socioculturally oriented, further reinforcing the appropriateness of adopting a focus group as the research method (Pearce & Lee, 2005). A focus group enables the simultaneous presentation of divergent and, occasionally, even conflicting opinions derived from the dynamic interactions within a group. The efficiency of focus group discussions is critically contingent on participation selection. Stewart and Shamdasani (1990) explained that the benefits and validity of focus group data will be affected by the extent to which participants can express their opinions in an open and comfortable manner. This statement highlights the importance of a relaxed environment for proper discussions and interactions among the panel members.

This study recruited participants from a class of Master's degree students in the tourism and hospitality field. All the participants were Chinese Mainlanders working in tourism-related businesses, such as hotels and travel agencies in Mainland China. All participants were part-time students working full-time. Most participants occupied managerial-level positions with extensive work experience. They flew to an eastern city in China a few times annually during their study period to attend block mode classes. Their salary was above the average level locally, while the tuition for this graduate degree program was over RMB 130,000, excluding other expenses for transportation and accommodation. All the participants had also traveled to Hong Kong to purchase luxury items, such as bags and watches, together with experiential products of luxury hotels. Therefore, including them is theoretically rigorous for exploring the fundamental motivations of traveling to Hong Kong, France, the US, and the United Kingdom (UK) for luxury items. Given that the participants are acquainted with one another via their Master's program, their identity as students is an advantage instead of a drawback in the study because such an acquaintance allows them to interact freely and discuss the issues comfortably. This interaction is vital for a focus group to generate an in-depth understanding of a phenomenon. In this manner, they could be regarded as people familiar with the conspicuous consumption of the Chinese and experienced with regard to Chinese perceptions of tourism activities. The study was conducted in Hangzhou when the participants gathered on campus to attend one of their study blocks. They were introduced to the project and invited to participate. After their consent was obtained, they were randomly assigned to five focus groups via lucky draw and their discussions were conducted in separate meeting rooms. A skilled trained moderator was assigned to each focus group to minimize subjectivity caused by having the same moderator. Each discussion lasted for an average of 70 minutes and the discussions were conducted in Chinese.

All discussions were recorded with the prior consent of the participants. The data were imported to NVivo 10 (QSR International, 2015), a software used in conducting content analysis on interview data. As a tool for storing, retrieving, categorizing, and coding text data, NVivo creates category trees to graphically present different categories simultaneously. Hence, NVivo has been extensively adopted by tourism and hospitality researchers in analyzing textual and visual materials (e.g. Pan, MacLaurin, & Crotts, 2007; Rodger, Moore, & Newsome, 2009). To ensure the validity of the data analysis, the researchers coded the content independently without prior discussions. The underlying motivations for luxury shopping and the advantages and challenges of Hong Kong as a luxury shopping destination were coded. When disagreements occurred in coding, the researchers discussed them until a consensus was reached. Upon completion of the data coding, results were compared, with similarities retained and differences resolved through discussions among the researchers. Similar to the data of other exploratory studies, those generated in the current research were not numerical. Instead, the data were in the form of protocols, patterns, and similarities that constituted a framework for linking the concepts in question.

Research context: Hong Kong as a destination for the Mainland Chinese

Located on the southern coast of China to the east of the Pearl River Estuary, Hong Kong has been well-known as a popular tourism destination in Asia for a long time. Since the early 1980s, the structure of the nature of inbound tourists to Hong Kong has changed and has been characterized by a considerable increase in the number of visitors from Mainland China (Zhang & Lam, 1999). The relaxation of travel regulations by the Chinese government has spurred a significant boom in the Chinese outbound market. Flows of tourists to Hong Kong have been impressive because of its geographical proximity, its introduction of the Individual Traveler's Scheme in August 2003, and the economic prosperity of China (Law, To, & Goh, 2008). In 2013, 75% of the total tourist arrivals in Hong Kong were from Mainland China, and this number is likely to increase in the future given the current growth rate (Hong Kong Tourism Board PartnerNet, 2015). Correspondingly, the importance of the Chinese market to the overall tourism industry in Hong Kong is evident.

Among the various tourism activities, shopping is repeatedly observed as an attractive factor for Chinese Mainlanders visiting Hong Kong (Huang & Hsu, 2005; Zhang & Lam, 1999). Scholars tend to use a typological (or cultural) lens to interpret and understand the shopping behavior of Chinese travelers. Mehta, Lalwani, and Ping (2001) observed that Asian consumers are less individualistic and their decision-making behaviors heavily rely on the extent to which they conform to cultural values. China is a collective society. Hence, the importance of social norms cannot be underestimated in the discussion of tourist behaviors. Hsu, Kang, and Lam (2006) argued that friends and relatives are the most influential factors in the decision-making of Mainland Chinese tourists. Similarly, Choi, Liu, Pang, and Chow (2008) suggested that Chinese Mainlanders belong to the group of ethnic-, art-, and people-oriented tourists who are heavily involved in purchasing crafts for themselves and as gifts. Pearce, Wu, and Osmond (2013, p. 149) credited the popularity among Chinese outbound tourists of maintaining a social linking behavior, whereby "bringing back gifts is the passport for re-entering the group after having been away."

Moreover, such gifts should be "distinctive and well-judged" (Pearce & Wu, 2012, p. 149).

Although Hong Kong has a solid basis for its development as a luxury shopping destination, this image has diminished in recent years because of the adverse effects of political and social movements in the local society. Such movements include the rising anti-Mainlander sentiment of the locals because of the inconvenience brought about by cross-border parallel trading and misbehaviors of Mainland travelers, as well as the increasing competitiveness of neighboring countries, such as Japan and South Korea. Therefore, understanding the luxury consumption behavior of Chinese Mainlanders and their view of Hong Kong as a luxury shopping destination is important in assisting Hong Kong to regain its reputation as a shopping paradise.

Findings and discussions

A total of 43 participants were included in the focus group discussions, with each round of discussions having eight to nine members. Of this number, over 60% are males aged 30–45 years. They have at least two years of work experience, with over 30% of them having over 10 years of work experience. Over 85% are from the industrial sectors of hotels and travel agencies, while 14% work in the tourism-related departments of universities or colleges.

Perceived parameters of luxury

Figure 1 presents the respondents' understanding on the concept of luxury and luxury items. The ranking of each parameter is based on the number of references made to it during the discussions. For most participants, luxury is closely related to brand and, in certain cases, brand is a proxy of luxury because other respondents used the terms interchangeably during the entire discussion. An analysis of the discussion results revealed that these brand-related attributes include brand stories and equity. This finding is expected because advertisers usually create specific meanings of advertised products in the modern business world that can eventually be interpreted and used by consumers in constructing their inner world (Elliott & Wattanasuwan, 1998). Despite the well-recognized brand effect, respondents in this study had limited knowledge of luxury brands. When brand stories were mentioned, the coordinators invited the respondents to further elaborate on these stories. Only few participants could provide specific examples. Similar findings have been obtained by previous studies, suggesting that Chinese luxury consumers tend to treat several best-known brands as representations of the entire category of luxury products (Atsmon & Dixit, 2009; Zhan & He, 2012).

Premium price was another important parameter in understanding luxury. Given the similarities to high price, this study coded words and terms of financial affordability to this category. This variable suggested the existence of the Veblen effect. Husic and Cicic (2009) explained that Veblenian consumers tend to impress others with prestigious prices, which they perceive to have a high conspicuous value. Price is typically an effective indicator of quality. That is, a high price indicates a high product or service quality. Moreover, certain brands (e.g. Hugo Boss) charge a higher price for products with smaller logos to entice customers who consciously or unconsciously refuse to serve as human billboards (Husic & Cicic, 2009). Hence, price in this case and the identified brand principally represent the perceived prestige and

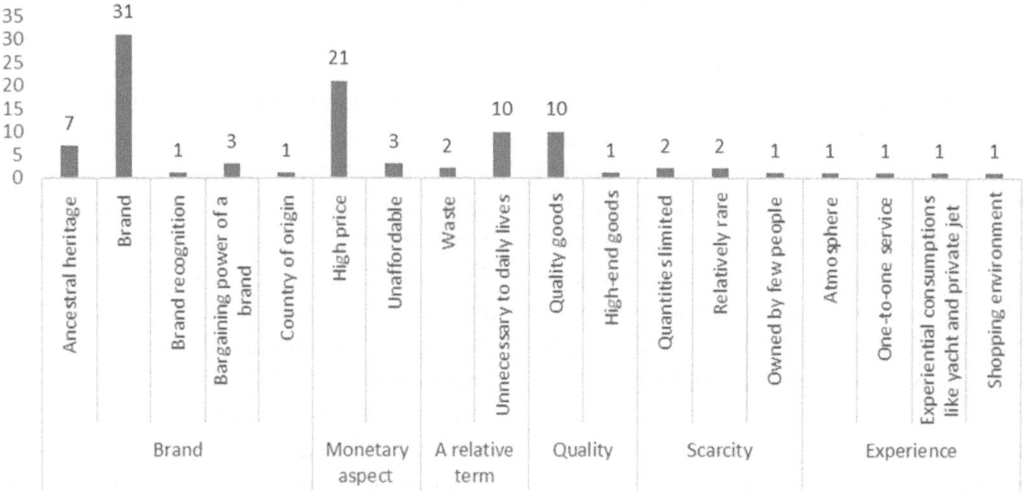

Figure 1. The concept of luxury: focus group findings.

superiority inherent in luxury consumption. The high ranking of the tangible attributes of brands and high prices likewise further emphasize that Chinese consumers chiefly rely on external factors rather than on their own judgment in luxury shopping.

The respondents also used a relative angle to express their understanding of luxury. They regarded the consumption of luxury goods as the purchase of unnecessary items relative to the necessities of daily life. This understanding indicates the relative perspective that is adopted in interpreting luxury because it depends on socio-economic status. This idea is consistent with the most-identified motivation for luxury consumption. Another relative view treats luxury as a "waste" relative to human materialistic needs. Although the term "waste" can mean a few unnecessary things in daily living, a distinctive line still exists between these two concepts. The "waste" opinion can be attributed to the traditional Chinese culture of thrift and frugality, which have been strongly cultivated among Chinese children for many generations (Wang & Lin, 2009). Correspondingly, buying unnecessary items can naturally be regarded as a waste.

Quality is another important parameter in defining luxury. As a key strategic ingredient, product quality has long been regarded as a fundamental element in determining business success (e.g. Jacobson & Aaker, 1987; Mathews-Lefebvre & Valette-Florence, 2014). Sheth, Newman, and Gross (1991) defined product quality as the functional aspect of commodities that refers to basic utilities and benefits. Although luxury goods are designed to satisfy the psychological and functional needs of consumers, the latter are typically diminished, which can be observed from the ratio of price to utility. Nueno and Quelch (1998) determined that the ratio of price to utility can be used in differentiating luxury brands from others, such as upmarket and premium brands. In particular, luxury brands have the highest price-to-quality ratio, thereby indicating the pursuit of product quality by certain respondents, who can be categorized as perfectionist consumers as proposed by Husic and Cicic (2009).

Rarity is another important parameter in defining luxury based on the focus group discussion. This parameter can be indicated by the phrases "quantities limited," "relatively rare," and "owned by few." The principle of rarity is substantially documented in the luxury consumption literature (Pantzalis, 1995). The exclusivity phenomenon can be achieved by several divergent mechanisms, such as setting a limited supply and high price, marketing tactics that have been frequently adopted by luxury marketers. This finding contrasts with the extant observations showing that rarity only holds in an individualistic culture (e.g. the US) instead of a collective culture (e.g. Singapore and Hong Kong) (Phau & Prendergast, 2000).

The respondents also emphasized the importance of experience in luxury consumption. A few of them expressed this opinion by comparing the shopping environment in China to that in European countries, whereas others compared the current environment of Hong Kong to how it was like in the past decade. Atwal and Williams (2009) observed the complicated process of luxury brand marketing that encompasses the tangible attributes reflecting quality and performance, and the experiential elements relating to consumer lifestyle. Hence, they recommended the marketing of the experiential aspects of luxury items because it is "about taking the essence of a product and amplifying it into a set of tangibles, physical, and interactive experiences that reinforce the offer" (Atwal and Williams, 2008, p. 341). This characteristic can be regarded as a feature of the desires of Chinese Mainlanders. This feature can also be attributed to the research context of luxury destination adopted in this study because experiential value has been extensively recognized as a key benefit in tourism and hospitality activities (Yuan & Wu, 2008).

Motivation for the involvement in luxury shopping

The identified motivations for luxury shopping are generally consistent with the preceding framework based on self-concept, and these motivations have likewise evolved along the spectrum ofsymbolism attached to the genuine pursuit of product quality (Tsai, 2005). As expected, the well-identified privileged images, such as premium price and prestigious names in luxury consumption, were also identified in this study. Moreover, this work ascertained several drivers with distinctive Chinese characteristics. Figure 2 presents the motivations identified in this study within the self-concept framework.

The motivations of the interdependent self are presented with six aspects: social conformity, *mianzi* (i.e. social status), vanity and bragging, dream value, gift-giving, and Western influences. Numerous studies have delineated the influences of social conformity in luxury consumption (e.g. Choo, Moon, & Kim, 2012). Chinese culture features a high level of social conformity, which has been regarded as a key predominant factor in analyzing publicly visible products, such as consumption (e.g. Park, Rabolt, & Sook Jeon, 2008). Social conformity fundamentally operates in our daily lives as reference group influences, which are further presented in the forms of informational and utilitarian influences.

Informational motivation functions on the basis of three perspectives: stories of certain brands, information

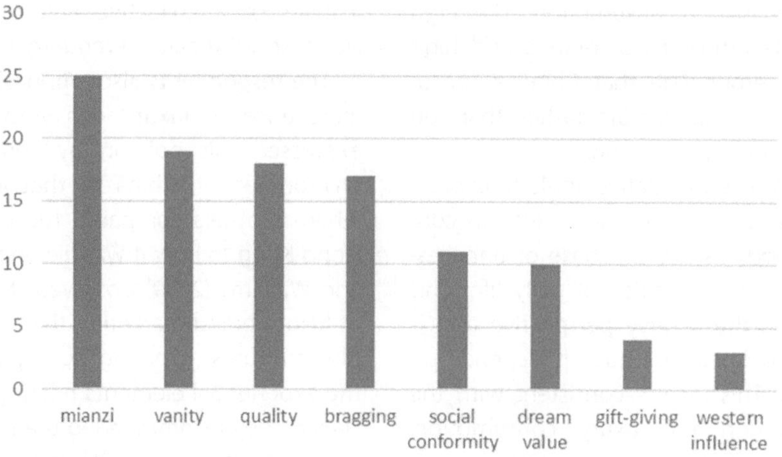

Figure 2. Motivations for luxury consumption.

of certain social groups, and information of the trends. Such informational motives are relatively superficial because the respondents claimed that many luxury buyers lack sufficient knowledge of luxury brands. In particular, these luxury buyers could list brand names while having limited knowledge of the specific meanings and themes inherent in the brands. This phenomenon partly contributed to the highly selected luxury brands when luxury consumption was empirically analyzed in China (e.g. Zhan & He, 2012), as well as to the popularity of luxury products that have publicly visible logos.

The utilitarian influence reflects the attempts of individuals to comply with the behavior of others to gain rewards or avoid punishments (Park & Lessig, 1977). This finding could be caused by the main consumer group in China. The focus group discussion indicated that the primary group of consumers of luxury goods in China comprises second-generation rich individuals. This result is supported by O'Cass and Choy (2008), who reported that Generation Y consumers (i.e. young adult consumers) have more positive attitudes toward premium brands and are more willing to pay additional amounts for these status brands. Hence, these customers would lose face in front of their peers if they lacked one or two pieces of luxury items. A respondent who works in a university used an extreme example to demonstrate this distorted phenomenon:

> Take some students from my university as an example. Some of them are poverty-stricken students, but their behaviors do not make them look like they are poor. They use iPhones, buy lots of snacks from supermarkets, and wear branded clothes ... They are regarded as poverty-stricken students, as their family incomes are below a certain level. However, they are willing to invest their limited money in things just because their classmates own them.

The second motivation for luxury consumption is *mianzi*, which is the Chinese version of social status and refers to "recognition by others of an individual's social standing and position" (Buckley, Clegg, & Tan, 2006, p. 276). Together with *guanxi, mianzi* is regarded as one of the core traditional Chinese values (Child & Lu, 1996). *Mianzi* is predominant in interpersonal relationships, and having *mianzi* means that people satisfactorily impress others with their reputation. Nearly all the respondents in the current study mentioned *mianzi* in their descriptions but in different forms. For example, a respondent referred to *mianzi* as follows: "This brand, this product, is well recognized by people in this society. Therefore, you will have face if you consume it." Another respondent suggested that *mianzi* "only if what you have is visible to others, either brands or logos, do you feel you have mianzi." Other respondents specifically defined the discussion scope by stating that

> in the whole society, those who were born in the 1980s or the 1990s are not that well paid. But if you live in a group in which all others are wearing branded clothes, if what you have is not branded, then there is a chance that you cannot settle in society and have no mianzi

.

Vanity and bragging are suggested as the third driver of luxury consumption, possibly relating to economic equality in Mainland China. Ordabayeva and Chandon (2011, p. 27) explained that inequality increases conspicuous consumption among bottom-tier consumers attempting to "keep up with the Joneses." This statement was echoed in parts of the discussion, such as in terms of examples or phrases, including "compare" and "vie against each other." A few respondents described this phenomenon as being applicable to the entire society by stating the following:

> We Chinese, more or less, have feelings of vanity. Take my elder sister as an example. My aunt is overweight, and one day she sent something to my sister at school. After she left, a classmate asked my sister, "Who is she?" My sister felt embarrassed and said, "Ah, my neighbor." If my aunt had been elegant and slim, maybe that would be a totally different story. Those young and middle-aged people involved in luxury consumption have the same psychological consideration.

For the Chinese the importance of face-saving behaviors and vanity is part of their very nature

However, a few respondents specifically identified women, stating that "Women like buying luxury items to show off."

The respondents also mentioned that Western influence and gift-giving can generate luxury consumption. Western influences can be interpreted on the basis of two perspectives: increasing exposure to Western products among the Chinese and the preference of the young generation for foreign values and products. Since the late 1980s, a rise in modernization has increased the exposure of China to Western products and services. A growing number of Chinese people, particularly the elite, had overseas experiences, which profoundly influenced their manner of thinking and consequently increased their acceptance of Western values. In the domestic market, multinational companies emphasized Western values in their advertisements, which successfully conditioned young Chinese to be more receptive to Western values and ideas (Zhang & Shavitt, 2003). The two views can be connected to the long-lasting worship of all Western things among the Chinese as well. Respondents stated that "The Chinese inherently have the worship for everything Western in their hearts," "It has something to do with the Chinese culture, which includes the Chinese consumption mentality," and "Worship for everything from the Western world could also contribute to the enlargement of the luxury market in China."

Gift-giving can also be understood on the basis of two perspectives. First, *guanxi*, which pertains to interpersonal relationships in China, is the cornerstone of the friendship or intimacy aimed at continued favor-seeking behaviors (Su & Littlefield, 2001). Hence, gift-giving is an essential means of maintaining interpersonal connections, although gift types may vary based on social obligations (i.e. the role of gift receivers to senders). For example, a respondent stated: We buy luxury products in bulk. What is that for? We buy it for gift-giving, as nowadays, items such as branded belts or wallets are the most concealed and appropriate gifts for officials."

Second, gift-giving can be another form of bragging by suggesting to their social circles that the gift-giver has been overseas.

Previous studies on luxury consumption conducted in the context of Western countries revealed a trend of diminishing symbolic and increasing personal values (Choo et al., 2012). Both trends were identified in the current work. Respondents regarded luxury items as a visual label, thereby impressing others with their own prominent achievement and enhancement of social status. Some respondents also claimed that they purchased luxury items for product quality.

Among those who stated the pursuit of quality, one respondent raised an interesting point by stating that "luxury goods are of good quality." Such a statement is similar to the observation of Husic and Cicic (2009, p. 242) that consumers in a post-socialist country "are very loyal to the idea that every well-known brand has to be of good quality." A few respondents suggested that "those middle-class people who buy luxury items for gift-giving and for self-use have a deep-rooted demand for quality." Another respondent expressed that "some people buy luxury items because they are aware of the brand and are loyal to it." Moreover, such enthusiasm for luxury items stems from the lack of confidence in the products of domestic brands. A respondent used the traveling suitcase as an example:

> I asked my friend in Germany to buy a Rimowa suitcase for me. It has a size of 20 inches and is made of an aluminum magnesium alloy. I asked my friend to buy it not necessarily because it is of superior quality, but because products of domestic brands are of low quality. I normally travel with a lot of stuff. There is a chance that the products of domestic brands would crash on the way, such as when the wheels break, which will cause me a lot of trouble.

Another respondent suggested that "if the quality of home-produced items is reliable, then we would still support made-in-China items."

Perceptions of Hong Kong as a luxury shopping destination

This study also analyzed the respondents' perceptions of Hong Kong as a luxury shopping destination. Table 1 presents the advantages and disadvantages of developing Hong Kong as a luxury shopping destination. Hong Kong, which is the most popular short-haul destination, has a solid basis for becoming a luxury shopping destination. Generally, the respondents claimed that they have adopted positive attitudes toward Hong Kong as a luxury shopping destination. However, Table 1 shows different results. The major strength of Hong Kong is its geographical proximity to Mainland China. Such a

Table 1. Advantages and disadvantages of Hong Kong as a luxury shopping destination.

Advantages	Disadvantages
• Geographical and cultural links (e.g. language, customs, and cultures)	• Inflation of prices in Hong Kong, such as that of accommodation
• Easy accessibility	• Emotional aversion caused by movements such as Occupy Central in 2014
• Well-developed facilities (e.g. developed transportation system) as well as good public security	• Unstable political and social environment
• Extensive product catalogues	• Decreasing service quality (e.g. hostile attitudes of locals and taxi drivers toward Chinese Mainlanders, incompatible travel behaviors and habits of Chinese Mainlanders, and crowded shopping environment)
• Advantageous interest rates and availability of UnionPay as a payment gateway	• Development of alternative shopping channels (e.g. other destinations for low transportation fees and easy visa liberalization in Mainland China such as the China (Shanghai) free trade zone and offshore duty policy in Hainan Province), and Daigou (i.e. overseas personal shoppers)

proximity enables it, as well as Macau, to be the traditional short-term overseas destination among Chinese Mainlanders. Cultural links, such as a low language barrier and a similar cultural background, are also beneficial in removing the travel constraints of Chinese Mainlanders to Hong Kong. Hong Kong also has a range of basic infrastructures, from the in-and-out transportation to the local in-city transportation system. Despite these unique advantages of Hong Kong as a tourism destination among Chinese Mainlanders, other merits appear to be marginally competitive in the market, including complete product catalogues that can represent competition for other fancy luxury destinations, such as Paris, London, and New York. Moreover, Hong Kong has good public security, attractive interest rates – which are closely related to the increasing competitiveness of the Renminbi in the international market – and the payment gateway UnionPay.

Despite such merits, Hong Kong faces an extremely challenging situation in attracting Chinese Mainlanders to shop for luxury items. These challenges are centered on emotional aversions caused by a few political and social movements, such as Occupy Central. These movements call for an enhanced democratic electoral system while giving Hong Kong an anti-Mainlander image in return. This situation was clearly illustrated through excerpts from the focus group discussions: "Nobody wants to go to places where residents are hostile to them" and "Hong Kong people now have prejudices against Chinese mainlanders." Before Hong Kong even recovered from the Occupy Central movement, the anti-parallel trading protest worsened the already declining tourism industry, particularly because gangster-like protesters occasionally attacked tourist-like passersby (*China Daily*, 2015).

The respondents also complained about the increasing price of accommodation, decreasing service quality, and unpleasant shopping environments partly caused by the increasing number of tourists and unfriendly attitudes of locals toward Chinese Mainlanders. Moreover, the emergence of alternative shopping channels further revised the market structure (e.g. daigou). Reduced transportation fees also substantially increased the cost efficiency of shopping in other destinations, such as South Korea and Japan. Visa liberalization further decreased the attractiveness of Hong Kong among Chinese Mainlanders because alternatives, such as Japan, South Korea, and Taiwan, became considerably more accessible. Hence, to secure an advantageous status, Hong Kong should adjust its destination policy and be innovative in designing and promoting tourism products.

Hong Kong, as a luxury shopping destination, was used to further corroborate the findings related to luxury definition and motivation. If the advantages and disadvantages presented in Table 1 are further sorted out, then the decreasing advantages of Hong Kong as a luxury shopping destination are chiefly related to its declining monetary efficiency. Political movements further intensifying the already declining situation are considered. A respondent in our study said:

> Residents from the first-tier cities would not regard Hong Kong as a luxury shopping destination because they have more options, like traveling to London and Paris where the promotions are more intensive. These locations have all the top-tier brands, whereas Hong Kong mainly has second-tier brands or even those brands that are only regarded as luxury brands by the Chinese, such as Coach. Those from second- or third-tier cities or counties may want to visit Hong Kong while shopping during the tour.

This statement implied the concept previously identified, thereby indicating that luxury consumption by Chinese Mainlanders is primarily driven by bragging that they have been overseas. After years of development, Hong Kong has become a mature destination among Chinese Mainlanders. For some Mainlanders, traveling to Hong Kong is very similar to domestic travel. This perception, which is not as appealing as two decades ago, began after Hong Kong's return to China in 1997. Given the marginal difference in prices between visiting Hong Kong and other countries, the majority of Chinese Mainland residents would naturally select the latter. Traveling overseas would make them feel that they have more *mianzi* than others.

Conclusions and implications

This study centers on the development of a luxury shopping destination. In particular, it focuses on three fundamental questions: What does luxury mean to the Chinese? What factors motivate them to indulge in luxury shopping? To what extent do perceptions of a specific place as a luxury shopping destination match consumers' understanding of luxury and motivations for luxury shopping? To answer these questions, this study conducted focus group discussions among Mainland Chinese residents. Increasing affluence has prompted the emergence of the Chinese as a new but appealing customer group in various business sectors. Luxury consumption is not exempt from this general trend. Insights into the behaviors of luxury consumers can be obtained only by considering the cultural differences in each unique society. Different characteristics inherent in the Chinese market require special attention and knowledge, thereby possibly further enriching the understanding of global mobility. Thrift and frugality are substantially valued in Chinese culture, as partly evidenced by the unequalled record of personal bank savings (Faure & Fang, 2008). Accordingly, luxury consumption has steadily gained momentum amid these traditional values. The preference of Chinese shoppers to buy luxury items overseas can be understood by considering the hefty import tariffs and taxes, as well as the strategy of premium prices in the Chinese market. Observations from the industry have confirmed that approximately two-thirds of luxury items are purchased outside of China (The Economist, 2015a).

The current findings are consistent with such a phenomenon, indicating the suitability of Hong Kong as the research context. The perceptions of luxury and shopping motivations of Chinese consumers can be simply but paradoxically summarized as follows: Chinese luxury consumers cling to thrift. The present study integrated this concept with the notion of Markus and Kitayama (1991) regarding the interdependent/independent self, and generally categorized the currently identified motives of Chinese consumers in pursuing luxury shopping (Figure 3). In summary, certain consumers may engage in luxury consumption to impress others with product attributes, such as price, exclusivity, and popularity. By contrast, those with an independent self purchase luxury products to fulfill their internal goals and needs.

To Chinese consumers, brand and premium price are relative surrogates of luxury and luxury items. Literature has long identified brand and price as indicators of high quality and value, which increase purchase willingness (Wiedmann, Hennigs, & Siebels, 2009). However, in the current study, price and brand were dominant factors in the decision-making process of luxury consumption. Results indicate that the respondents are not equipped with the backstories of the luxury brands. Moreover, their purchase decisions are not driven by the congruence between self and brand image. Instead, their judgment to buy luxury items mainly relies on the popularity and price of brands, thereby possibly simplifying the process of choice selection and reducing any related risks. This finding may offer insights into consumers' conscious purchasing of counterfeit luxury brands in China. Brand prestige lifestyle immensely influences consumers who only have experience with counterfeit luxury fashion brands (Li et al., 2012; Phau & Teah, 2009).

Chinese Mainlanders buy to impress others. Thus, the dominance of conspicuousness underlies the phenomenon of "the Chinese speak the international language of shopping." This phenomenon was serially reported by *The New York Times*, 2006. The presentation of social status, which is a key element of conspicuous consumption, among the Chinese is *mianzi*. In a Confucian-oriented society, the concept of *mianzi* (face) is of paramount importance in China (Lindridge & Wang, 2008). The belief that "face is the spiritual creed of the Chinese people" has deeply penetrated the normal lives of the Chinese. This belief is reflected in several proverbs, such as "People live on their face, trees live on their skin" (Shi, Furukawa, & Jin, 2011. p. 163). To Chinese consumers, luxury items (i.e. branded and high-priced items) carry symbolic meanings. The four types of motivations identified in the current study point toward *mianzi* issues. Brands (or simply logos) and premium prices relatively carry the meaning of rarity and high social status, thereby making Chinese Mainlanders feel superior and have *mianzi*.

Our findings also suggest that complete product category and advantageous interest rates are critical in assessing if a place is a luxury shopping destination. Such results are consistent with the economic value and prestige brand involved in luxury consumption. The inherent mobility nature of tourism is also well implied as geographical distance, with easy accessibility highlighted as a critical merit of Hong Kong as a luxury shopping destination. This discovery parallels the aforementioned significance of economic value relating to

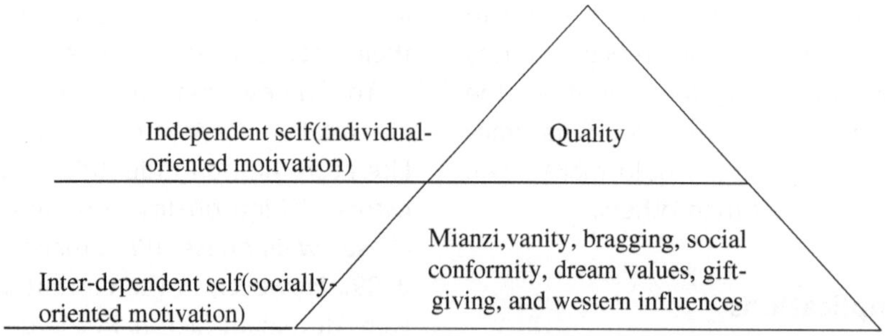

Figure 3. Interpretations of self-concept in understanding luxury consumption among the Chinese.

luxury consumption. This study also discovered the crucial role of consumer experience in developing luxury shopping destinations. This development is simultaneously inherent in the nature of the tourism industry because of the tangible nature of competitive sources within it (Stamboulis & Skayannis, 2003) and of luxury consumption because the experience is considered when luxury brand marketing is involved (Atwal & Williams, 2009).

This study likewise identified conscious halo effects in developing luxury shopping destinations. Previous studies suggested that the halo effect is "a fundamental inability to resist the affective influence of global evaluation on the evaluation of specific attributes, especially when they are unaware of the halo's existence" (Nisbett & Wilson, 1977, p. 255). In the current study, the respondents were well aware of the edge of going to Hong Kong to purchase luxury items. However, they consciously tolerated such edges to be diluted by halo incidents, such as the Occupy Central movement. Other political incidents may cause social chaos, thereby striking the tourism industry. However, the political and social movements in Hong Kong affect the Mainland Chinese, and the appeals of such movements include those related to national secession. The denial of national sovereignty goes against the patriotic and nationalistic beliefs of many Chinese people.

Theoretically, this study enriched the interpretations of self-concept theory in understanding Chinese luxury consumption motivations. It answers the question raised by Lau (1996) whether a concept that focuses on Chinese children and adolescents exists in the Chinese culture. Lau (1996) refuted the existence of self in the Chinese culture because it focuses considerably on the social identities of individuals. By contrast, the current study indicated that the socially oriented self dominates in Mainland China. The reason is that luxury consumption behaviors chiefly involve shopping to impress others and gain *mianzi* (i.e. social status). An increasing number of Chinese customers have recognized their inner self considerably because the drivers of their involvement in luxury consumption are quality items meeting their personal tastes, as well as their pursuit of intangible luxury experiences. The current study also contributes to destination image literature by recognizing the existence of halo effects. Moreover, the influence of factors, such as those social movements causing halo effects, are more imperative compared with the stimuli of luxury shopping. Therefore, future studies on destination image should judiciously analyze objects of interest within a place and assess their interrelationships before going on to inspecting a generally homogenous destination.

Given the diversified demands in the Chinese market, luxury retailers and shopping destinations (e.g. Hong Kong) should adopt differentiated marketing strategies. Promotions and advertising campaigns that create brand awareness should continue. Moreover, certain luxury customers from China are on a strict budget, as vividly illustrated by The Economist (2015a): "Chinese tourists have no problem buying Prada by day but sleeping in two-star hotels by night." Hence, second-hand luxury shops or outlets could attract Chinese shoppers. In addition, buying to show off is rare in high-end markets because shoppers genuinely pursue quality and shopping experiences. Highly discrete and customized products and services would be appealing to this segment. Hence, luxury brands in this market could win the game by competing in activities that evoke rarity, brand awareness, and quality.

Furthermore, a certain group of customers genuinely pursues quality. On the one hand, such customers have low confidence in the quality of domestic brand items. On the other hand, their taste, which seeks quality, can only be fulfilled by internationally well-known brands. This phenomenon presents business opportunities for luxury retailers; those who follow their feelings and personal tastes can substantially regulate their behaviors (Wong & Ahuvia, 1998). This finding implies the possibility of cultivating loyal customers in the Chinese market. However, this group is quality-oriented; it flocks to home-made brands if the quality of such brands could be ensured because these customers do not blindly seek luxury items. Given the halo effects, destination management agencies in Hong Kong should collaborate with others who play an important role in attracting luxury consumers from Mainland China and their dominant source market in the long run.

The limitations of this study are mainly related to the study sample. The participants in this study were representatives of luxury consumers. They offered in-depth views on luxury consumption motivations and perceptions of Hong Kong as a luxury shopping destination. Nevertheless, their similar, comparable backgrounds likely hinder divergent points of view. Despite the benefits of focus groups in qualitative research inquiries, a few respondents may hesitate to share their personal spending or luxurious experience in public. Further investigations of the issue by using other methods, such as in-depth or photo interviews, will be beneficial to understanding this phenomenon further.

Funding

This work was supported by The Hong Kong Polytechnic University [Project No. PolyU 155031/17B].

References

Ab Karim, S., & Chi, C. G. Q. (2010). Culinary tourism as a destination attraction: An empirical examination of destinations' food image. *Journal of Hospitality Marketing & Management, 19*(6), 531–555.

Albayrak, T., Caber, M., & Çömen, N. (2016). Tourist shopping: The relationships among shopping attributes, shopping value, and behavioral intention. *Tourism Management Perspectives, 18*, 98–106.

Alden, D. L., He, Y., & Chen, Q. (2010). Service recommendations and customer evaluations in the international marketplace: Cultural and situational contingencies. *Journal of Business Research, 63*(1), 38–44.

Atsmon, Y., & Dixit, V. (2009). Understanding China's wealthy. *The McKinsey Quarterly*, 4, 32–33.

Atwal, G., & Williams, A. (2009). Luxury brand marketing–the experience is everything! *Journal of Brand Management, 16* (5–6), 338–346.

Bain & Company. (2015). Luxury goods worldwide market study Fall-Winter 2014: the rise of the borderless consumer. Retrived June 17, 2015, from http://freesky.azurewebsites.net/browse.php?u=6W6Py0uz9DAQEKaazMXnfJ0VRJ7lChJv1PU3Ncqxjs1sVX1JG8XGXlyZPFB2MsPwXb1%2BMPJBnOcZ08QWMx9MojmbGK1yNKp8ghqjA0l%2BOqwbtFQ1xJt4co7KQvE%3D&b=0.

Beerli, A., Meneses, G. D., & Gil, S. M. (2007). Self-congruity and destination choice. *Annals of Tourism Research, 34*(3), 571–587.

Berthon, P., Pitt, L., Parent, M., & Berthon, J. P. (2009). Aesthetics and Ephemerality. *California Management Review, 52*(1), 45–66.

Beverland, M. (2004). Uncovering "theories-in-use": building luxury wine brands. *European Journal of Marketing, 38*(3/4), 446–466.

Buckley, P. J., Clegg, J., & Tan, H. (2006). Cultural awareness in knowledge transfer to China-The role of guanxi and mianzi. *Journal of World Business, 41*(3), 275–288.

Calder, B. J. (1977). Focus groups and the nature of qualitative marketing research. *Journal of Marketing Research, 14*(13), 353–364.

Chen, A., & Peng, N. (2014). Examining Chinese consumers' luxury hotel staying behavior. *International Journal of Hospitality Management, 39*, 53–56.

Child, J., & Lu, Y. (1996). Introduction: China and international enterprise. In J. Child & L. Yuan (Eds.), *Management issues in China: international enterprises*, (pp. 1-18). London: Routledge.

Childers, T. L., & Rao, A. R. (1992). The influence of familial and peer-based reference groups on consumer decisions. *Journal of Consumer Research, 19*(2), 198–211.

China Daily. (2015). Hong Kong protesters attacking tourists are worse than gangsters. Accessed on July 18, 2015 via http://www.chinadaily.com.cn/opinion/2015-03/11/content_19775660.html.

Choi, T. M., Liu, S. C., Pang, K. M., & Chow, P. S. (2008). Shopping behaviors of individual tourists from the Chinese Mainland to Hong Kong. Tourism Management, 29(4),811–820.

Choi, S., Liu, L., & Kim, D. Y. (2015). Accessing Tourists' Unconscious Associations about International Destinations: Data Fuzzification of Reaction Times in the Implicit Association Test. *Journal of Travel & Tourism Marketing, 32*(5), 578–594.

Chon, K. S. (1992). Self-image/destination image congruity. *Annals of Tourism Research, 19*(2), 360–363.

Choo, H. J., Moon, H., & Kim, H. (2012). Luxury customer value. *Journal of Fashion Marketing and Management: An International Journal, 16*(1), 81–101.

Christodoulides, G., Michaelidou, N., & Li, C. H. (2009). Measuring perceived brand luxury: an evaluation of the BLI scale. *Journal of Brand Management, 16*(5), 395–405.

Correia, A., Kozak, M., & Kim, S. (2017). Investigation of Luxury Values in Shopping Tourism Using a Fuzzy-Set Approach. Journal of Travel Research. doi:10.1177/0047287517741005

Correia, A., Kozak, M., & Kim, S. (2018). Luxury shopping orientations of mainland Chinese tourists in Hong Kong: Their shopping destination. *Tourism Economics, 24*(1), 92–108.

Dimanche, F., & Samdahl, D. (1994). Leisure as symbolic consumption: A conceptualization and prospectus for future research. *Leisure Sciences, 16*(2), 119–129.

Ekinci, Y., Sirakaya-Turk, E., & Preciado, S. (2013). Symbolic consumption of tourism destination brands. *Journal of Business Research, 66*(6), 711–718.

Elliott, R., & Wattanasuwan, K. (1998). Brands as symbolic resources for the construction of identity. *International Journal of Advertising, 17*(2), 131–144.

Eng, T. Y., & Bogaert, J. (2010). Psychological and cultural insights into consumption of luxury western brands in India. *Journal of Customer Behaviour, 9*(1), 55–75.

Esu, B. B., & Arrey, V. M. E. (2009). Branding cultural festival as a destination attraction: a case study of Calabar Carnival Festival. *International Business Research*, *2*(3), 182–192.

Li, F., & Chris, R. (2018). Souvenir shopping experiences: A case study of Chinese tourists in North Korea. *Tourism Management*, *64*, 142–153.

Faure, G. O., & Fang, T. (2008). Changing Chinese values: Keeping up with paradoxes. *International Business Review*, *17*(2), 194–207.

Goody, J. (2006). From misery to luxury. Social Science Information, 45(3),341–348.

The Free Dictionary. (2015). Luxury. Retrived June 15, 2015, from http://www.thefreedictionary.com/luxury.

Han, Y. J., Nunes, J. C., & Drèze, X. (2010). Signaling status with luxury goods: The role of brand prominence. Journal of marketing, 74(4),15–30.

Heath, A. P., & Scott, D. (1998). The self-concept and image congruence hypothesis: An empirical evaluation in the motor vehicle market. *European Journal of Marketing*, *32* (11/12), 1110–1123.

Hennink, M., Hutter, I., & Bailey, A. (2011). *Qualitative Research Methods*. London: Sage Publications Ltd.

Heung, V. C., & Qu, H. (1998). Tourism shopping and its contributions to Hong Kong. *Tourism Management*, *19*(4), 383–386.

Hong Kong Tourism Board PartnerNet. (2015). A statistical review of Hong Kong tourism 2013. Retrived June 15, 2015, from http://securepartnernet.hktb.com/filemanager/intranet/dept_info/private_20/paper/Stat-Review/Stat_Review2013/stat_review_2013_n_2.pdf.

Hsu, C. H., Kang, S. K., & Lam, T. (2006). Reference group influences among Chinese travelers. *Journal of Travel Research*, *44*(4), 474–484.

Huang, S., & Hsu, C. H. (2005). Mainland Chinese residents' perceptions and motivations of visiting Hong Kong: Evidence from focus group interviews. *Asia Pacific Journal of Tourism Research*, *10*(2), 191–205.

Hung, K., Guillet, B. D., & Zhang, H. Q. (2018). Understanding Luxury Shopping Destination Preference Using Conjoint Analysis and Traditional Item-Based Measurement. *Journal of Travel Research*. doi:10.1177/0047287518760259

Hung, K., & Petrick, J. F. (2011). The role of self and functional congruity in cruising intentions. *Journal of Travel Research*, 50(1), 100–112.

Husic, M., & Cicic, M. (2009). Luxury consumption factors. *Journal of Fashion Marketing and Management: An International Journal*, *13*(2), 231–245.

Hwang, J., & Han, H. (2014). Examining strategies for maximizing and utilizing brand prestige in the luxury cruise industry. *Tourism Management*, *40*, 244–259.

Jacobson, R., & Aaker, D. A. (1987). The strategic role of product quality. *The Journal of Marketing*, *51*(4), 31–44.

Jansen-Verbeke, M. (1990). Leisure+ shopping= tourism product mix. In G. Ashworth & B. Goodall (Eds.). *Marketing tourism places* (pp. 128–137). London, UK: Routledge.

Jap, W. (2010). Confucius face culture on Chinese consumer consumption values toward Global Brands. *Journal of International Management Studies*, *5*(1), 183–192.

Jin, H., Moscardo, G., & Murphy, L. (2017). Making sense of tourist shopping research: A critical review. *Tourism Management*, *62*, 120–134.

Kapferer, J. N., & Bastien, V. (2009). The specificity of luxury management: Turning marketing upside down. Journal of Brand Management, 16(5–6), 311–322.

Kastanakis, M. N., & Balabanis, G. (2012). Between the mass and the class: Antecedents of the "bandwagon" luxury consumption behavior. *Journal of Business Research*, *65* (10), 1399–1407.

Lau, S. (1996). Self-concept development: Is there a concept of self in Chinese culture. In S. Lau (Eds). Growing up in the Chinese Way: Chinese Child and Adolescent Development (pp. 357-374). Sha Tin, Hong Kong: The Chinese University Press.

Law, R., To, T., & Goh, C. (2008). How do Mainland Chinese travelers choose restaurants in Hong Kong? An exploratory study of individual visit scheme travelers and packaged travelers. *International Journal of Hospitality Management*, *27*(3), 346–354.

Lee, J. H., & Hwang, J. (2011). Luxury marketing: The influences of psychological and demographic characteristics on attitudes toward luxury restaurants. *International Journal of Hospitality Management*, *30*(3), 658–669.

Li, F, & Chris, R. (2018). Souvenir shopping experiences: A case study of Chinese tourists in North Korea. Tourism Management, 64, 142–153.

Li, G., Li, G., & Kambele, Z. (2012). Luxury fashion brand consumers in China: Perceived value, fashion lifestyle, and willingness to pay. *Journal of Business Research*, *65* (10), 1516–1522.

Li, J. J., & Su, C. (2007). How face influences consumption. *International Journal of Market Research*, *49*(2), 237–256.

Lindridge, A. M., & Wang, C. (2008). Saving "face" in China: Modernization, parental pressure, and plastic surgery. *Journal of Consumer Behaviour*, *7*(6), 496–508.

Litvin, S. W., Goldsmith, R. E., & Pan, B. (2008). Electronic word-of-mouth in hospitality and tourism management. *Tourism Management*, *29*(3), 458–468.

Markus, H. R., & Kitayama, S. (1991). Cultural variation in the self-concept. In J. Strauss & G. R. Goethals (Eds.), *Multidisciplinary perspectives on the self* (pp. 18–48). New York: Springer-Verlag.

Mathews-Lefebvre, C., & Valette-Florence, P. (2014). Manufacturer brand value and the respective role of brand sensitivity, situational involvement and enduring involvement. *Journal of Brand Management*, *21*(3), 236–253.

Megehee, C. M., & Spake, D. F. (2012). Consumer enactments of archetypes using luxury brands. *Journal of Business Research*, *65*(10), 1434–1442.

Mehta, S. C., Lalwani, A. K., & Ping, L. (2001). Reference group influence and perceived risk in services among working women in Singapore: A replication and extension. *Journal of International Consumer Marketing*, *14*(1), 43–65.

Moscardo, G. (2004). Shopping as a destination attraction: An empirical examination of the role of shopping in tourists' destination choice and experience. *Journal of Vacation Marketing*, *10*(4), 294–307.

Mussalam, G. Q., & Tajeddini, K. (2016). Tourism in Switzerland: How perceptions of place attributes for short and long holiday can influence destination choice. *Journal of Hospitality and Tourism Management*, *26*, 18–26.

Nisbett, R. E., & Wilson, T. D. (1977). The halo effect: Evidence for unconscious alteration of judgments. *Journal of Personality and Social Psychology*, *35*(4), 250.

Nueno, J. L., & Quelch, J. A. (1998). The mass marketing of luxury. *Business Horizons, 41*(6), 61–68.

O'Cass, A., & Choy, E. (2008). Studying Chinese generation Y consumers' involvement in fashion clothing and perceived brand status. *Journal of Product & Brand Management, 17*(5), 341–352.

Ordabayeva, N., & Chandon, P. (2011). Getting ahead of the Joneses: When equality increases conspicuous consumption among bottom-tier consumers. *Journal of Consumer Research, 38*(1), 27–41.

Pan, B., MacLaurin, T., & Crotts, J. C. (2007). Travel blogs and the implications for destination marketing. *Journal of Travel Research, 46*(1), 35–45.

Pantzalis, I. (1995). Exclusivity strategies in pricing and brand extension. Retrived June 21, 2015, from http://arizona.openrepository.com/arizona/bitstream/10150/187351/1/azu_td_9620410_sip1_m.pdf.

Park, C. W., & Lessig, V. P. (1977). Students and housewives: Differences in susceptibility to reference group influence. Journal of consumer Research, 4(2),102–110.

Park, H. J., Rabolt, N. J., & Sook Jeon, K. (2008). Purchasing global luxury brands among young Korean consumers. *Journal of Fashion Marketing and Management: An International Journal, 12*(2), 244–259.

Park, K. S., Reisinger, Y., & Noh, E. H. (2010). Luxury shopping in tourism. *International Journal of Tourism Research, 12*(2), 164–178.

Pearce, P. L., & Lee, U. I. (2005). Developing the travel career approach to tourist motivation. *Journal of Travel Research, 43*(3), 226–237.

Pearce, P. L., Wu, M. Y., & Osmond, A. (2013). Puzzles in Understanding Chinese Tourist Behavior: Towards a Triple-C Gaze. *Tourism Recreation Research, 38*(2), 145–157.

Phau, I., & Prendergast, G. (2000). Consuming luxury brands: the relevance of the 'rarity principle'. *The Journal of Brand Management, 8*(2), 122–138.

Phau, I., & Teah, M. (2009). Devil wears (counterfeit) Prada: a study of antecedents and outcomes of attitudes towards counterfeits of luxury brands. *Journal of Consumer Marketing, 26*(1), 15–27.

Podoshen, J. S., Li, L., & Zhang, J. (2011). Materialism and conspicuous consumption in China: A cross-cultural examination. *International Journal of Consumer Studies, 35*(1), 17–25.

QSR International. (2015). Nvivo 10 for Windows. Retrived May 25, 2015, from http://www.qsrinternational.com/products_nvivo.aspx.

Rabbiosi, C. (2015). Renewing a historical legacy: Tourism, leisure shopping and urban branding in Paris. *Cities, 42*, 195–203.

Rodger, K., Moore, S. A., & Newsome, D. (2009). Wildlife tourism, science and actor network theory. *Annals of Tourism Research, 36*(4), 645–666.

Sheth, J. N., Newman, B. I., & Gross, B. L. (1991). Why we buy what we buy: a theory of consumption values. *Journal of Business Research, 22*(2), 159–170.

Shi, Z., Furukawa, I., & Jin, C. (2011). Cognitive discrepancy in Chinese "face": Mian and lian, and their impact on cognition of country-of-origin image. *Frontiers of Business Research in China, 5*(2), 163–178.

Sirakaya-Turk, E., Ekinci, Y., & Martin, D. (2015). The efficacy of shopping value in predicting destination loyalty. *Journal of Business Research, 68*(9), 1878–1885.

Sirgy, M. J. (1982). Self-concept in consumer behavior: A critical review. *Journal of Consumer Research, 9*(3), 287–300.

Sirgy, M. J., & Su, C. (2000). Destination image, self-congruity, and travel behavior: Toward an integrative model. *Journal of Travel Research, 38*(4), 340–352.

Smith, V. L. (Ed.). (2012). Hosts and guests: The anthropology of tourism. Philadelphia, Pennsylvania: University of Pennsylvania Press.

Spencer-Rodgers, J., Boucher, H. C., Mori, S. C., Wang, L., & Peng, K. (2009). The dialectical self-concept: Contradiction, change, and holism in East Asian cultures. *Personality and Social Psychology Bulletin, 35*(1), 29–44.

Stamboulis, Y., & Skayannis, P. (2003). Innovation strategies and technology for experience-based tourism. *Tourism Management, 24*(1), 35–43.

Stewart, D. W., & Shamdasani, P. N. (1990). Focus Groups: Theory and Practice. London, UK: Sage Publications.

Su, C., & Littlefield, J. E. (2001). Entering guanxi: a business ethical dilemma in mainland China? *Journal of Business Ethics, 33*(3), 199–210.

Swanson, K. (2014). Souvenirs, tourists, and tourism. In A. A. Lew, C. M. Hall, & A. M. Williams (Eds.), *The Wiley Blackwell Companion to Tourism* (pp. 179–188). Chichester, UK: John Wiley & Sons, Ltd.

The Economist. (2015a). Beyond bling: tastes are changing, but appetites remain keen. Retrived June 15, 2015, from http://www.economist.com/news/special-report/21635763-tastes-are-changing-appetites-remain-keen-beyond-bling.

The New York Times. (2006). Chinese Speak the International Language of Shopping. Accessed on June 18, 2018 via https://www.nytimes.com/2006/11/07/world/europe/07eurochinese.html.

Timothy, D. J. (2005). *Shopping tourism, retailing and leisure*. Clevedon, USA: Channel View Publications.

Todd, S. (2001). Self-concept: a tourism application. *Journal of Consumer Behaviour, 1*(2), 184–196.

Truong, Y., & McColl, R. (2011). Intrinsic motivations, self-esteem, and luxury goods consumption. *Journal of Retailing and Consumer Services, 18*(6), 555–561.

Tsai, S. P. (2005). Impact of personal orientation on luxury-brand purchase value. *International Journal of Market Research, 47*(4), 429–454.

Vickers, J. S., & Renand, F. (2003). The marketing of luxury goods: An exploratory study–three conceptual dimensions. *The Marketing Review, 3*(4), 459–478.

Vigneron, F., & Johnson, L. W. (1999). A review and a conceptual framework of prestige-seeking consumer behaviors. *Academy of Marketing Science Review, 1*(1), 1–15.

Wang, C. L., & Lin, X. (2009). Migration of Chinese consumption values: traditions, modernization, and cultural renaissance. *Journal of Business Ethics, 88*(3), 399–409.

Wiedmann, K. P., Hennigs, N., & Siebels, A. (2009). Value-based segmentation of luxury consumption behavior. *Psychology & Marketing, 26*(7), 625–651.

Wong, I. A., & Wan, Y. K. P. (2013). A systematic approach to scale development in tourist shopping satisfaction: Linking destination attributes and shopping experience. *Journal of Travel Research*, *52*(1), 29–41.

Wong, N. Y., & Ahuvia, A. C. (1998). Personal taste and family face: Luxury consumption in Confucian and Western societies. *Psychology and Marketing*, *15*(5), 423–441.

Wu, M. Y., & Pearce, P. L. (2012). Tourism research in and about Tibet: Employing a system for reviewing regional tourism studies. Tourism and Hospitality Research, 12(2),59–72.

Yeung, S., Wong, J., & Ko, E. (2004). Preferred shopping destination: Hong Kong versus Singapore. *International Journal of Tourism Research*, *6*(2), 85–96.

Yuan, Y. H., & Wu, C. K. (2008). Relationships among experiential marketing, experiential value, and customer satisfaction. *Journal of Hospitality & Tourism Research*, *32* (3), 387–410.

Zhan, L., & He, Y. (2012). Understanding luxury consumption in China: Consumer perceptions of best-known brands. *Journal of Business Research*, *65*(10), 1452–1460.

Zhang, B., & Kim, J. H. (2013). Luxury fashion consumption in China: Factors affecting attitude and purchase intent. *Journal of Retailing and Consumer Services*, *20*(1), 68–79.

Zhang, J., & Shavitt, S. (2003). Cultural Values in Advertisements to the Chinese X-Generation–Promoting Modernity and Individualism. *Journal of advertising*, *32*(1), 23–33.

Zhang, Q. H., & Lam, T. (1999). An analysis of Mainland Chinese visitors' motivations to visit Hong Kong. *Tourism Management*, *20*(5), 587–594.

"It is a way of life": detecting Chinese students' wellbeing during the Spring Festival homecoming

Ying Zhao, Fangfang Chen and Honggang Xu

ABSTRACT

This paper follows the call for a "mobility turn" in the social sciences and attempts to understand the relationship between the annual Spring Festival homecoming and wellbeing by focusing on student mobility. Sixteen students are selected; their movements for two weeks around the Spring Festival are tracked, and their experiences are solicited in interviews. The relationship of interest is explored in three dimensions: physical movement, social connections and contextual shifts. The results show that students are eager to pursue an ideal home for their wellbeing but do not always obtain positive wellbeing. In reality, they face both physical and mental contradictions.

Introduction

Eighteen days before the 2018 Spring Festival, while many migrants across China were gradually returning to their hometowns, an article about a graduate of Peking University who had refused to return home for twelve years appeared on the Sina homepage. Wang Meng, a top student based on college entrance exams, complained that his parents were too controlling when he was growing up. An unpleasant experience during the Spring Festival twelve years ago made him refuse to return home in the following years. This experience represents the same sad story for thousands of families who long for their students to return, but it also reveals the contradictions between joy and worry during the Spring Festival homecoming.

Student mobility is significant, but it has attracted little attention in tourism marketing research in contemporary China. The country's rapid urbanization has led to people frequently moving to different areas, especially to the large cities. Rural-urban floating labourers often find a temporary dwelling in the city and are willing to ultimately return to their hometown (Huang, King, & Suntikul, 2017). Meanwhile, highly educated migrants are likely to build a stable residence in a city other than their hometown (Cui, Geertman, & Hooimeijer, 2015) but return to their hometown regularly to see their parents and relatives. University students in large cities are in midst of education-to-work transitions (Cairns, 2008) and are therefore different from the previous two groups. Mobility is heightened during the transition to adulthood, as it overlaps with rapid urbanization and cultural individualization (Zhao, Dijst, & Chai, 2016). For students, "home" still refers to their original hometown. However, some of them might be aware that they will not return to their hometowns for work but will settle in other cities in the near future.

As an important student mobility behaviour, homecoming helps us understand the process of personal transition. University students are in the period of "becoming" rather than "being" adults. They are attached to their original home but are far away in distance. They may begin to prepare to leave home for a long time to live an independent life, even after middle school or high school studies. However, their time at university is when they are expected to prepare for their independence financially, socially, and psychologically. Closeness and alienation exist simultaneously in their heart, but they feel tension and anxiety when they return home on holiday. University students attempt to pursue good feelings at home to obtain the sense of belonging that is lacking in cities. It seems that emotional attachment to one's hometown, linked to childhood memories, exists deep in the heart regardless of the number of years that have passed (Tuan, 2007). However, they may actually end up with mixed feelings, including anxiety about the uncertain future and conflict between their imagination and the reality of their hometown. Wellbeing is considered an effective concept to measure the extent of feeling in harmony for a happy life (Ryan & Deci, 2001). It is widely used in mobility and tourism studies to understand tourist experiences (Kou, Xu, & Hannam, 2017; Li & Chan, 2017);

thus, we adopted this concept to explore how students experience being back home when they are on holiday.

Although students' homecoming in China is a significant phenomenon during the Spring Festival, few studies have examined this issue. This paper focuses on student mobility during the Spring Festival and its relationship with student wellbeing. We selected five graduate students at a university in southern China as empirical cases and tracked their movements and experiences for two weeks (one week before the New Year and one week after the New Year in 2017) as the first-stage survey. In addition, eleven other students were interviewed regarding their experiences during their Spring Festival homecoming visits as the second-stage survey to test and complete the findings. By applying the "therapeutic mobilities" concept (Gatrell, 2013), we attempt to illustrate the wellbeing that students obtain on their journey. This paper is intended to demonstrate the usefulness of "therapeutic mobilities" in mobility and wellbeing for exploring the personal growth experience of current Chinese students in a China that is rapidly transitioning.

Literature review

Visiting friends and relatives (VFR)

Students' homecoming shares some similarities with visiting friends and relatives (VFR), such as visiting friends and relatives who are close in relation but distant in space, facing intensified social interaction in a short term and feeling a sense of obligations, etc. Traditionally, VFR is discussed in the tourism literature, and the value of VFR as a worthy form of tourism is often emphasized (Griffin, 2013; Janta, Cohen, & Williams, 2015). However, VFR actually constitutes the intersection of migration, mobility, and tourism (Janta et al., 2015); it is more central to the human experience than leisure tourism (Seaton, 2017). Thus, the complex social relationships, the meaning of place, and the affirmations and obligations involved with this phenomenon should be explored (Janta et al., 2015; Seaton, 2017). In accordance with this call, a number of studies, mostly qualitative ones, have emerged in recent years to illustrate human experience.

In these recent studies, VFR is understood as a constellation of diverse practices that may include social relationships, the provision of care, affirmations of identities and roots, the maintenance of territorial rights, and leisure; in addition, VFR often occurs on special occasions such as rituals and yearly events (Janta et al., 2015). The meaning and effects of these practices are demonstrated by analysing the experience of visiting (e.g. Huang et al., 2017; Seaton, 2017). In most cases, VFR is associated with pleasant experiences and positive emotions, including a sense of continuity and connection with the past, family closeness, and authentic tourism experiences (Huang et al., 2017; Seaton, 2017). However, sometimes there might also be tension and disappointment. For example, the geographic or social distance between visitors and the people they visit, language bias, unrealistic expectations, and an unsatisfactory environment may cause awkwardness and unpleasant feelings between visitors and the visited (Huang et al., 2017; Seaton, 2017).

As the most basic VFR event (Seaton, 2017), the homecoming visit shares certain characteristics with general VFR. Among studies that shed light on the meaning and features of the homecoming visit, Pearce's (2012) work on visiting home and familiar places (VHFP) is the most comprehensive. According to Pearce (2012), VHFP involves going home or going to locations marked by "individuals' experience during their period of growing up" (Pearce, 2012, p. 1024). Pearce (2012) also notes that a sense of wellbeing might be enhanced in VHFP by stimulating positive emotions, positive memories, and a positive attitude towards the past and future. However, as noted in VFR studies, the negative aspects of homecoming should not be neglected. A mismatch between the imagined and experienced homecoming might decrease wellbeing (Stefansson, 2004), and the sources of this mismatch are similar to those in VFR studies (e.g. the social distance between hosts and visitors, a different living environment, cultural differences) (Markowitz & Stefansson, 2004).

College students' homecoming visits during the Spring Festival largely fall into the category of VHFP because they are visits to the places where they grew up. However, they differ from a general homecoming visit for the following reasons: 1) they are performed during a festival for a reunion, and the motives to visit and the activities engaged in during the visit to some extent involve obligations. 2) The students do not have a "home" in the cities in which they reside; thus, the home that they annually visit is their present home, not their past home. 3) The home to which they are attached is now gradually being alienated. They clearly understand that they are in the process of preparing to set up their own homes in cities far away from this home. Thus, the happiness, disappointment, contradictions and wellbeing that they experience during the homecoming visit could be different from those in general VFR or VHFP studies. It will be interesting to investigate the effect of this homecoming visit on their wellbeing.

Homecoming during a transitional stage

The homecoming that is the focus of this paper occurs during a transitional stage for students. College students are in the transition to adulthood, a life stage that can be referred to as emerging adulthood (Arnett, 1994, 2007). This life stage extends from their late teens to their mid- to late 20s, in which they are emerging into adulthood but are not there yet (Arnett, 2007).

College students' transitional period is characterized by increased freedom, increased exposure to new ideas and increased concern about privacy (Arnett, 1994; Stephenson-Abetz & Holman, 2012). However, they do not yet recognize themselves as independent adults (Arnett, 1994). They have moved out of their parents' house but mostly live in dormitories, where some of their daily arrangements are taken care of by adults (Arnett, 1994). They have left their family connections and old relationships spatially behind but maintain them through online communication; they also try to negotiate their old connections with their families and their new relationships with friends (Stephenson-Abetz & Holman, 2012). They are connected to "home" in a unique way, such that both connections and tensions exist (Stephenson-Abetz & Holman, 2012). The homecoming visit is where these connections and tensions intensify.

In addition, the characteristics of emerging adulthood are not universal across cultures (Arnett, 2007); in fact, studies have identified numerous culture-specific features (Fuligni, 2007; Nelson, Badger, & Wu, 2004; Seiter & Nelson, 2011). The concept of emerging adulthood was first proposed based on Western industrialized societies and reflected Western individualism (Arnett, 1994, 2007). Researchers suggested that students from families or societies with an Asian background possessed a strong sense of family obligation that reflects collective values (Fuligni, 2007; Nelson et al., 2004; Seiter & Nelson, 2011). Family responsibility and Confucian doctrine play an important role in Chinese college students' emerging adulthood behaviourally, cognitively and emotionally (Nelson et al., 2004). Insights into college students' experience during the Spring Festival homecoming visit will benefit our understanding of the tension they face in emerging adulthood.

China is also experiencing a transitional and dynamic process. The country's rapid urbanization has made people frequently move to different areas (Chan, Liu, & Yang, 1999). For some individuals, their childhoods involved a process of living in several places, as their family moved from one dwelling place to another. Other individuals may have experienced stable childhoods but relocate to pursue educational or employment opportunities (Cui et al., 2015). Among these groups, a high proportion of students have been shown to relocate when they enter university (He, 2014). In addition, due to limitations on education resources, the concentration of high-quality educational resources in large cities intensifies students' movement and further affects their family life and school life. In the past two or three decades, Chinese students have experienced a process of gradually leaving home. This process has created a separation between students' dwelling cites and their hometowns, such that home and the places that people come from have new meanings, and returning home also has comprehensive meanings that might be related to wellbeing. Although students leave home and pursue higher education in another city, they continue to regularly return home and enjoy family life (Li, 2014; Pearce, 2012).

Homecoming during the Spring Festival is a fixed and secular journey for Chinese students. The festival is a common time to return home for reunions with one's extended family. Thousands of students or workers return to their hometowns before the Spring Festival, which causes extremely high traffic loads on national transportation systems, called *Chunyun*. *Chunyun* means the annual migration from urban to rural areas and from east to west during the Spring Festival in China, which is understood as a "national ritual" (Li, 2014). Rapid modernization and urban development in China have caused an unprecedented migration of rural labourers to cities. The homecoming during the Spring Festival puts tremendous pressure on railways, roads and even airline traffic systems. In this situation, buying a ticket is indeed a challenge for the students. With regard to the "homecoming ritual", the students share similarities with migrant workers.

This type of behaviour can promote students' wellbeing because the home to which they return may host a sense of belonging that is absent in students' everyday life in the city (Li, 2014), and it might have positive effects on their physical, mental and social wellbeing that they need deep in their heart. As Pearce (2012) notes, VHFP may involve complex feelings and emotions. However, VHFP is not always positive; it is sometimes negative. In addition, returning home during the Spring Festival to some extent is an obligation and a ritual performance. Students do not always obtain positive wellbeing when they return home; the trip can be a psychological, physical and financial challenge. Therefore, some migratory students who refer to themselves as the "home-fear group" (Li, 2014) question the meaning of returning home.

Mobility and wellbeing

Wellbeing is a holistic concept that describes the "optimal psychological functioning and experience" (Ryan & Deci, 2001, p. 142) of human beings, and it is complex and heterogeneous in nature. Theorists from diverse disciplines (e.g. psychology, sociology, economics, geography) have developed a variety of definitions and explanations of wellbeing. Among the various yet interrelated understandings of wellbeing, the distinction between the hedonic approach (Kahneman, Diener, & Schwarz, 1999) and the eudaimonic approach (Waterman, 1993) has been widely discussed. Hedonic conceptions understand wellbeing as pleasure and happiness, while eudaimonic conceptions consider that wellbeing lies in the realization of human potential. To resolve this controversy, Ryan and Deci (2001) suggest that wellbeing should be understood as a multidimensional phenomenon that includes insights from both the hedonic and eudaimonic approaches. A number of studies have proposed dimensions of wellbeing that have both hedonic and eudaimonic elements, for example, good physical and mental health, a strong social network, a secure income and home, self-continuity, and self-fulfilment (Kou et al., 2017; Ziegler & Schwanen, 2011).

With the rise of the mobility turn (Urry, 2007), the connection between mobility and wellbeing has attracted the attention of a number of scholars. This connection lies in the fact that when moving around, people are exposed to factors that influence their wellbeing (Kwan & Schwanen, 2016). Empirical studies have found that various aspects of mobility, such as daily activities, daily travel, modes of transportation, and seasonal migration, will affect an individual's wellbeing (Freedman, Carr, Cornman, & Lucas, 2017; Jones, Harms, & Heinen, 2016; Kou et al., 2017; Stanley, Hensher, Stanley, & Vella-Brodrick, 2011; Ziegler & Schwanen, 2011).

In 2013, Gatrell published a paper that proposed the concept of "therapeutic mobilities" to address the importance of mobility for health and wellbeing. He highlighted the therapeutic qualities of the act of moving from one place to another and some of the wellbeing gains from the mobilities or journeys that people take (Gatrell, 2011). He explored the relationship between walking, as one form of mobility, and wellbeing. Walking contributes to wellbeing in the following aspects: 1) walking as a physical activity improves people's physical and mental health; 2) walking helps people to be socially connected; 3) walking connects people with the contextual environment. Later, Kou et al. (2017) developed the framework by applying it to explore the relationship between elders' seasonal migration and wellbeing (Kou et al., 2017; Kou, Xu, & Kwan, 2018). They explored how seasonal migrant retirees in Sanya tried to maintain wellbeing by setting up daily life routines, establishing social connections, and interacting with the surrounding context (Kou et al., 2017). In their latest work, they further claimed that retirees experience wellbeing through processes of constructing physical, social and emotional aspects in their origin and destination places (Kou et al., 2018).

In Kou et al.'s (2017) work, wellbeing was constructed in five dimensions: physical wellbeing, mental wellbeing, social wellbeing, self-continuity and self-fulfilment. Physical wellbeing refers to the condition of physical health, which is related to perceptions of adequate rest, a life with a slow pace and one's daily routine. Mental wellbeing refers to the perception of psychological status. Social wellbeing refers to the support and communication that mobile people receive through their social relations. Self-continuity refers to the maintenance of a sense of familiarity and belonging and a continuation of people's previous life. Self-fulfilment refers to the spirit of pursuing the good life and self-development through the homecoming experience.

In our study, we adopted the analytical framework of Kou et al. (2017) for the following reasons: 1) their work examines daily activities within the seasonal migration context. 2) College students' homecoming is understood as a way for them to find themselves in their transition to adulthood, in which the dimensions of self-continuity and self-fulfilment are especially relevant. 3) The framework was proposed and tested in the Chinese context. Thus, it is understood that college students' daily mobility during the Spring Festival homecoming visit will influence their wellbeing in these five dimensions and might include 1) their daily activities and movement, 2) the social connections that they experienced, and 3) the environmental context with which they interacted while being mobile.

Research design

Data collection

This paper adopted a two-stage procedure for data collection. The first and main stage involved five participants and followed time-geographical theory. As a powerful method to measure mobility (Hägerstrand, 1970), time-geographical theory has recently developed 3D visualizations of students' space-time trajectories collected by GPS to measure migrants' long-distance mobility (Frändberg, 2008, 2014) and the day-to-day variability in their mobility (Shen, Kwan, & Chai, 2013). Attached to space-time tracking, momentary feelings or emotions can be recorded through the experience sampling method (Shoval, Schvimer, & Tamir, 2018), which uses several simple questions to measure the emotions triggered by a location transformation or a fixed time interval.

The data in the first stage were collected from 21 January to February 2 2017. The survey period covered one week before and one week after the New Year. The data collection involved three parts. First, each student was given a mobile phone installed with *Happy Tracker*, a mobile APP that records spatial movements every 30 seconds. Second, the participants were asked to report the 1) personal location meaning (own home, alma mater, etc.); 2) activities, 3) emotion (happy, neutral or unhappy), 4) companion, 5) text describing the moment and 6) pictures (optional) every two hours via another app called *Path*. The required times for posts were 10:00, 12:00, 14:00, 16:00, 18:00, 20:00, and 22:00; they were also free to post at other times. An alarm was set in each phone to remind the participants to post on *Path*. Third, after the students returned to university, they were interviewed independently about their return trip home, their everyday life and social contacts at home, and any perceived problems or mental satisfaction during the Spring Festival homecoming. The interviews lasted from 20 to 60 minutes. All interviews were recorded with permission and transcribed later.

Strict criteria were followed to ensure the reliability and validity of the data in this process. First, smart phones were installed with the two applications, and the necessary alarms for the activity reports were set on the tracking devices. Second, respondents at the same university were recruited. This university was chosen because it offered easy access for the survey team. Additionally, the university guaranteed that the students' training would be finished before the formal survey and that the devices would be returned after the survey. Third, local natives were initially excluded. Some students who planned to travel together with their families were also excluded. We selected five students who were highly interested and who volunteered as final participants because the tracking survey lasted for two weeks, and each respondent needed to have sufficient patience to report every two hours. We finally obtained 394 valid activity records. Logically, the number of activity records should equal 455, as calculated from 5 students × 7 times per day × 14 days. There are fewer valid records than estimated for two reasons. One is the occurrence of weak signals in rural areas, which made posting online impossible. The other is that participants sometimes forgot to post on time even when an alarm was set.

For the first stage of data collection, the authors performed an initial analysis. First, GPS tracking data were visualized into space-time paths in ArcScene 3D GIS software. The participants' space-time paths reflected their mobility patterns, which provided background information on their daily activities in their hometowns. Then, the records from *Path* were carefully sorted and analysed descriptively, including information on activities, companions, and emotions. This analysis provided a brief idea of the most frequently conducted activities, the intensity of the participants' social connections during the Spring Festival, and their general emotional states. Analysis of variance (ANOVA) provided information about the participants' different emotional states when conducting different activities or in the presence of different groups of companions. Next, transcripts of interview and text information from *Path* were analysed using thematic analysis. Thematic analysis is an effective method in qualitative data analysis; it "is a method for identifying, analysing and reporting patterns (themes) within data" (Braun & Clarke, 2006, p. 79). The analytical process in this study followed four steps: 1) the transcript and text were read repeatedly by the author to allow immersion in the data; 2) the initial coding was conducted manually, and all potential patterns were coded; 3) the initial codes were written on a separate paper and reorganized into theme piles; 4) the themes were reviewed and refined to ensure that they were supported by the data (Braun & Clarke, 2006). Finally, some detected themes emerged: "mobility", "social connections", "mobile context" and "wellbeing".

To validate the initial results, in-depth interviews were conducted as the second stage after the Spring Festival in 2018. The interviewees were recruited from the same university following the same criteria. Eleven students were interviewed by recalling their experiences during the Spring Festival homecoming, and the interviews lasted from 25 to 65 minutes. Data were also analysed by thematic analysis. The results further verified themes developed in the first stage. Triangle validation by resampling and complementary information helped to ensure the results and conclusions.

Sample case context

To understand the distribution of the dwelling cities and hometowns of the sample students, we summarized their long-distance journey behaviours and corresponding experiences. The hometowns of the both first-stage participants (S01-S05) and second-stage participants (S06-S16) varied in spatial distribution. The linear distance from home to school varied from 82.5 to 2250 kilometres, as shown in Table 1. For the homecoming, each student had to overcome at least two obstacles: obtaining a ticket in *Chunyun* and undertaking and enduring a difficult journey.

The journey home was difficult for most of the students. High-speed trains were the main transport mode for the long-distance trip home (Table 1). However, all of the journeys required transferring among different types of transportation in the departure or destination cities, which caused the length of their journeys to be relatively long, varying from 3 to 48 hours. For example, Sample 02, who comes from Sichuan, needed to take a coach to Guangzhou, a train from Guangzhou to Chongqing, and a coach from Chongqing to Mianyang. Although the train journey lasted 20 hours, the transfer time brought the total length of the trip to nearly two days. Another example is Sample 05, whose home was the closest. She had to take

Table 1. General information on the homecoming trips.

No.	Hometown	Distance (km)	Transport Mode	Time Distance (h)	Family information
First Stage					
S01	Yiyang, Hunan	701	Taxi→High-speed train1→High-speed train2→Car	7	Lives with parents
S02	Mianyang, Sichuan	1345	Coach→Subway→Train→Coach	30	Lives with parents
S03	Lushan, Jiangxi	837	Coach→Subway→Train→Coach	13	Lives with parents and one brother
S04	Yicheng, Hubei	954	High-speed train	6	Lives with parents and has three sisters who live in Wuhan
S05	Taishan, Guangdong	82.5	Bus→Coach→Bus	3	Lives with parents and one brother
Second Stage					
S06	Anqing, Anhui	980	Coach→Airplane→High-speed train→Coach	12	Lives with parents and one sister
S07	Panjing, Liaoning	2250	Coach→Subway→Train→Car	30	Lives with parents
S08	Handan, Hebei	1615	High-speed train	10	Lives with parents
S09	Xianning, Hubei	850	Coach→Train→Car	15	Lives with parents and one brother
S10	Nanning, Guangxi	550	High-speed train	6	Lives with parents and one brother
S11	Pingtan, Fujian	730	Coach→Trian→High-speed train→Coach	17	Lives with parents
S12	Jingmen, Hubei	950	Self-driving (with parents)	2 days	Lives with parents in Guangdong Province; will return to Hubei Province to celebrate Spring Festival
S13	Yingtan, Jiangxi	750	Coach→Airplane→Car	7	Lives with parents
S14	Conghua, Guangdong	150	Couch	4	Lives with parents
S15	Jinan, Shandong	1600	Coach→Airplane→Car	8	Lives with parents and one brother
S16	Linyi, Shandong	1500	Coach→Trian→Car	30	Lives with parents

a coach from Zhuhai to Taishan and a public bus in each city. During her three-hour journey, she had to continuously pay attention to transfer stops and leave enough time to wait for the coach. An extreme example is Sample 12. He drove to hometown in Hubei Province for nearly two days because of congestions in highways. In addition to the cost in time, the unpleasant journey was a physical struggle, and the crowded trains and emotional tensions caused complicated feelings among the sample students. The students in the survey finally arrived at their hometowns approximately one week before the Spring Festival.

Results

Decision to return home

Approaching the Spring Festival, the decision whether to go home comes first. In some ways, home is constructed in terms of family and/or social relations (Liu, 2014). Seeing one's parents, relatives and friends is the most important reason for returning home because *"Spring Festival is the most typical festival for reunion (S16); I must go back home and see those people; there's a bond between us (S13)"*. Returning home for the Spring Festival was special for this student because it is a long-expected opportunity for reunion in a place where his/her family is staying or used to stay during a festival designed for reunion. In addition, the rituals that are performed for the Spring Festival intensify the atmosphere of reunion. These rituals may include cleaning, lighting firecrackers, writing poetry, worshiping ancestors, preparing the New Year's Eve dinner, and visiting friends. As mentioned by S04 and S14, *"I would clean, paste couplets on the gate, worship, and prepare for New Year's Eve dinner with my parents ... All of these things are all done together by the family, and it creates an atmosphere of being at home and Spring Festival"*.

The second question is where to go because the concept of home is multiple for most students. As shown in Table 1, the home to which students return is often not limited to the place where their parents live now but to the multiple places they lived while growing up. Unlike the older generation, who lived in the places where they were born throughout their lives, university students born in the late 1980s and 1990s in China are characterized by high mobility. They might have experienced several migrations and relocations in their early life due to the labour migration of their parents, the pursuit of better education or for other reasons (Fang, Yang, & Meiyan, 2009; Seeborg, Jin, & Zhu, 2000). Take S04 as an example. He was born in a small town in Hubei province in central China. The first relocation came when he was in primary school. He moved to a nearby city with his parents and lived there until the end of high school. After his college entrance exams, he went one thousand kilometres away to pursue his higher education in Guangxi province in southwestern China. Currently, he is pursuing a postgraduate degree in Guangdong province in southern China. He is not the only case. Ten of our 16 samples relocated to another town or city with their families before going to college. The process of relocation involves not only separation but also building connections. Memories and emotions are attached to specific places in this process and to multiple homes.

According to Pearce (2012), "home" refers to "the constellation of locations and their associations which individuals experience during their period of growing up" (p.1024). For college students who have grown up in a highly mobile environment, there could be multiple places that they call "home". Based on the experiences of our participants, this constellation includes the "old home" (*laojia*, where the students spent their childhood), the "(present) home" (where their parents live now and their dwelling space when returning home), and the "alma mater" (where they used to study). According to the data from the first stage of the research, four participants mentioned the tradition of visiting the "old home" with their parents, and three mentioned visiting the alma mater in their notes on *PATH*. Thus, returning home involved not only the house they called home in small cities or towns but also the "old home" and the alma mater, all of which they find to be emotionally significant. These multiple homes and the meanings and relations that they carried were the reasons for the students' homecoming.

Home usually seems to be associated with positive emotions, but alternative associations with tension and unfamiliarity should not be neglected. Tensions could be triggered by a changed body clock, intensified social interactions, the generational gap and student's feeling of uncertainties in a transitional stage. In the following section, we will use Kou et al.'s framework to examine homecoming and student wellbeing.

Physical movement

Compared with other parts of the year, the Spring Festival is characterized by intensified physical movement and activities. Numerous rituals and traditions are required during the Spring Festival, as mentioned by S01: *"There is a must-meet list; several friends from high school are on this list. I also need to visit my grandma in the old home"*. Physical movement is involved in most cases. Physical movement and activities can be better tracked by apps like *Happy Tracker* or *Path* than by interviews, thus analysis in this part is mainly based on data collected from the first stage (5 samples). To better understand the students' physical movement and activities, we separated their movements based on the spatial scope involved: first, activity and mobility around home (small-scale mobility) were analysed; then, we moved to large-scale mobility (mobility beyond the space they called "home").

Small-scale mobility

Daily activities conducted at home are important to understand the participants' physical and mental wellbeing. Here, "home" refers to the house in which their parents live and the neighbourhood, usually specified as "home" or "around home" by the participants when writing notes on *PATH*. They generally spent quite a long time at home. For example, both S01 and S02 spent nearly five full days at home (never stepping out of their dwelling spaces) during the research period. Therefore, in this section, we first analyse the daily activities at home.

The activity information is drawn from the records in *PATH*. The respondents were required to keep an activity diary that reported both their objective location (e.g. apartment, street, district) and subjective location (e.g. my own home, a friend's home, recreational places) every two hours. From 394 valid activity records, we selected those with the subjective location of "home" or "around home" and the emotional status as valid records. This process excluded 140 records for a final total of 254 activities. The purposes of these activities include personal affairs (30.7%), leisure activities (25.6%), work (22.4%), family affairs (13.8%) and others (7.5%).

The rhythm of conducting one's personal affairs was reflected in the students' "body clock". Personal affairs included sleeping, dining, and personal care. In most cases, the students needed to adjust their body clocks to be in line with the rhythm of their family. On the one hand, the life pace at home was usually slower, easier and more relaxed compared with that at university. On the other hand, festival traditions with fixed rituals at fixed times might disrupt their previous rhythm. Getting up early, staying up late and intensified social activities are common during festival days, especially on Chinese New Year's Eve, Chinese New Year's Day, and two days after Chinese New Year's Day in the Chinese lunar calendar.

The changed body clock had multiple effects on the students' wellbeing. The slow life pace may promote physical and mental wellbeing by ensuring sufficient rest. As mentioned by most respondents, they adjusted their body clock quite quickly and smoothly, enjoying the slow life pace at home: *"When I'm at home, my mom will ask me to go to bed early ... I actually like it because it guarantees enough sleep and helps me be energetic the next day" (S03)*. However, when the slow life pace and changed body clock encountered work pressure from school, frustration and tension emerged, negatively influencing physical wellbeing, mental wellbeing, self-continuity and self-fulfilment. Take S02 and S05 as an example: *"I was always sleepy at home, and it influenced my work efficiency. I felt frustrated when I couldn't finish the work on time" (S02);* mental wellbeing, self-continuity and self-fulfilment decreased because the changed body clock decreased the student's work efficiency. *"I want to sleep late if I stayed up late to work the night before, but my parents will wake me up to have breakfast, so I can't get enough sleep" (S05);* physical wellbeing decreased because of the different time arrangements of family members. Additionally, the intense and irregular arrangements during festival days may decrease the students' mental and physical wellbeing because they put extra pressure on the body and mind: *"I was unhappy on New Year's Day (1 January on the lunar calendar) because I stayed up late to welcome the new year and was pulled out of bed by my mom to go worship in the early morning" (S02).*

Work is an important type of activity, even during the Spring Festival, with both positive and negative effects on wellbeing. Work takes place through the wide use of smart phones. The students stayed connected with their university during the entire holiday. The students brought some assignments, works in progress, and papers to be written to their hometown. In the survey, the students' tensions and contradictions were mainly due to low efficiency. In addition to the changed body clock mentioned above, family responsibilities also influenced the energy devoted to work. As mentioned by S02, *"You felt you need to help your parents clean the house, purchase festival materials, visit relatives ... I'm not sure if it's because your parents expected you to do so, or because you enjoy it, or you just want to find an excuse to escape from work".* Conflicts erupted when a task was delayed, or guilt and self-accusation were aroused, thus influencing the students' mental wellbeing and self-fulfilment. However, if a balance between work and the holiday was successfully reached, then there was a mutually reinforcing effect between self-fulfilment from finishing work and joy from the festival celebration, which led to enormous joy and happiness. As experienced by S01, *"It was the night before New Year's Eve when I finished the work and emailed it. Then, I could set myself free and enjoy the festival ... I've never been so happy before, even during the Spring Festival".* The happiness expressed here is a mixture of mental health, self-continuity and self-fulfilment.

Leisure activities and family affairs were significantly associated with positive emotions. We examined five types of activities and emotions expressed by the participants using cross-tabulation and found significant differences ($\chi^2(8) = 54.06$, $p < .001$), as shown in Table 2. Leisure activities included reading, watching television/movies, playing cards, and hanging around. Family affairs included doing housework and watching children. Among 65 leisure activities, the "happy" emotion was expressed 58.5% of the time. In addition, for 35 family affairs, the "happy" emotion was reported 48.6% of the time. Family reunion is the main theme of the Spring Festival; thus, the students participated extensively in chatting, watching TV, and cleaning the house with family members. The positive emotions in these activities benefited their physical, mental and social wellbeing.

Large-scale mobility

Activities and travels beyond the students' present hometown occurred frequently during the Spring Festival. We use a space-time path in a time geographic framework to find the patterns of spatial distribution and flexibility in Figure 1. Activities beyond the students' present hometown revealed a significant connection with the old home, *laojia* in Chinese, where they used to live in their childhood. As shown in Figure 2, S01 travelled between the home and old home on January 25, 27, 28, 29 and 30, which were the most important days of the Spring Festival. The respondents showed a special connection to their old homes through their physical movements. Their emotional connections will be elaborated upon in the next two sections.

Travels beyond the present hometown were also shown to be long distance and irregular trips that carried positive emotions. Different from S01, S02 covered a larger spatial scale, with several trips to her uncle's home, as shown in Figure 2. S05 spent one day with family members at a hot spring resort in another city, feeling "happy" the whole day. During the driving time on long trips, three students mentioned happiness being together with their core family members. For example, S01 posted a picture of her father's back at 8:11 a.m. and described a lovely scene of a family trip: *"We're driving to my aunt's home. I'm trying to take a photo of my father, so you know that it's not my fault I have such a big face. But he just playfully turned away. I feel happy".* Long-distance trips provided a chance to enjoy scenery or a situation different from the students' everyday circumstances and an opportunity to communicate with family members in

Table 2. Emotions expressed in leisure and family affairs activities.

Emotion	Total sample*	Leisure and amusement activities	Family affairs	Personal affairs	Working	Others	F	Df	Sig.
Happy	35%	58.5%	48.6%	20.5%	8.8%	68.4%	54.06	8	.000
Neutral	58.7%	38.5%	42.9%	70.5%	84.2%	31.6%			
Unhappy	6.3%	3.1%	8.6%	9.0%	7.0%	0%			
Valid number	254	65	35	78	57	19			

Notes: *notes posted at home or near home were selected; the chi-square test was adopted.

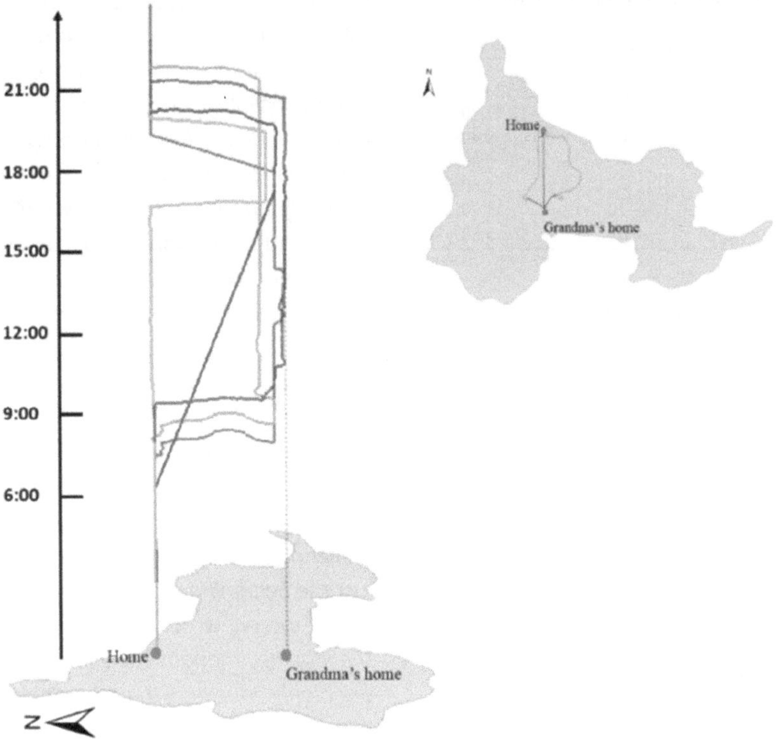

Figure 1. Temporal-spatial path of Sample A.

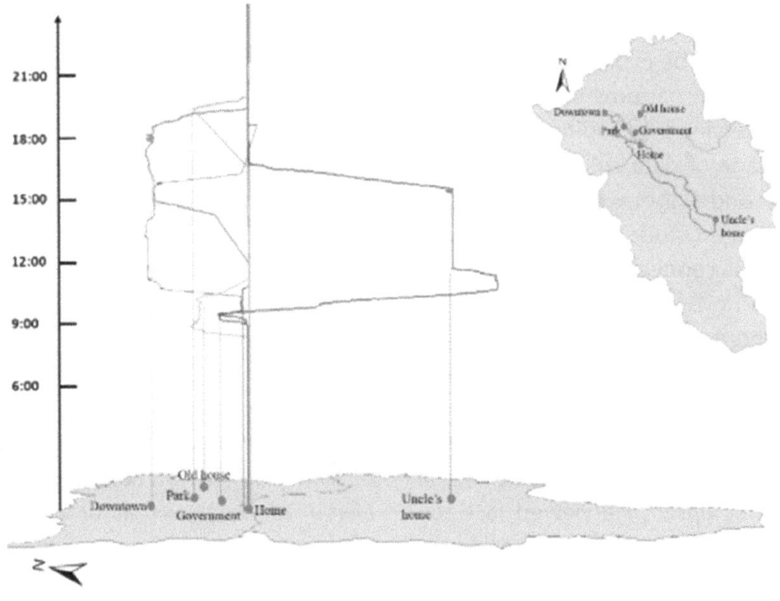

Figure 2. Temporal-spatial path of Sample B.

a mobile context, and these trips further contributed to the students' mental and social wellbeing.

Social connections

Intensified social interaction is one of the most important features of homecoming during the Spring Festival. Evidence was found about social interaction in the information about activity companionship. In the first stage of data collection, the five participants were required to provide information in *PATH* about the people with whom they engaged in activities. The results on companionship are shown in Table 3. First, the activities were conducted alone in only 33% of the situations, indicating that the

Table 3. Sharing of companionship in the Spring Festival.

Companionship role	Percentage	No. of companions	Percentage
Alone	33.00%	0	33.00%
Parents	40.60%	1	13.50%
Friends	25.90%	2	17.30%
Relatives		3	9.40%
-Young generation	19.80%	4	5.30%
-Same generation	25.90%	5	4.80%
-Older generation	15.00%	6–10	11.20%
		>10	5.50%

students were accompanied in most cases. The activities were mostly related to personal affairs such as sleeping, dining, or working. Second, regarding the number of companions, 36.2% of the situations included three or more companions. These two data points mean that the students encountered intensified social connections during the Spring Festival. Third, parents were the most frequently chosen companions, accounting for 40.6% of the situations. Brothers and sisters appeared in 25.9% of the situations. This result indicates that core family members played an important role in the respondents' wellbeing during the Spring Festival, providing both mental and physical support. This aspect will be further elaborated on below. Fourth, older-generation family members in extended families, such as aunts, uncles and grandparents, appeared in 15.0% of the situations, which was the lowest share compared with young-generation family members at 19.8% and same-generation family members at 25.9%. The causes of this low share will be explained below.

After obtaining the general patterns of companionship, information from interviews (including both the 5 samples from the first stage of data collection and the 11 samples from the second stage) were used to capture the relationship between students' social connections and wellbeing. Social connections might benefit mental and social wellbeing by providing an intimate and supportive atmosphere. S04 described the large family scene as follows: "*(The happiest thing is) having New Year's Eve dinner with the family, my parents and sisters; we haven't met up for a long time … Relatives are also on the must-meet list; you'll count: aunt, uncle … if someone is missed, you'll be upset*". He viewed face-to-face meetings as necessary for family members who were close in relation but far in distance. "*There's a bond between us, and I felt a sense of happiness from this bond (S13)*". As migrating student, they found the sense of belonging when staying with family members.

However, social interactions can also be a source of pressure and tension. Home, especially for a large family or extended family in China, is understood to be a haven in one's personal life (Duyvendak, 2011). The implication is that home is always open to blood relations and that people at home can live and communicate without reservation (Zhao et al., 2016). However, in most cases, the sense of intimacy may decrease because of long-time separation or different lifestyles. For the students, visiting elderly family members and reunions with extended families become a responsibility and are not enjoyable. They may feel bored "*I didn't feel emotionally connected to them (the extended family) and didn't know what to say to them … It's really boring and … awkward*" (S12). In extreme cases, stress and frustration can occur. Core family and extended family members often have no awareness of personal privacy. In the survey, the older generation was concerned about issues such as work, salary, romantic relationships or marriage of the young generation. These concerns were considered signs of care by the older generation but bothered the young generation. S05 expressed her depression in one note posted when chatting with her large family: "*(being with) aunts and uncles discussing relatives' education, career or marriage in public, I think you understand … but it's just the beginning … mentally defeated and tired, unhappy*". In the interview conducted later, she added, "*one thing is the job; they thought you should get a great job because you are highly educated; the other thing is the boyfriend … actually, I don't have an answer, I'm not sure about these issues. It's annoying when someone keeps asking you something about which you are not sure*". As a result of decades of high mobility in Chinese society, the young generation normally leaves their hometown at a very young age, and they are separated from their extended family both spatially and mentally. This separation in living space and the gap in worldviews caused tensions between the young and old generations, decreasing the students' mental wellbeing, social wellbeing and self-continuity.

Most of the time, being home with the core family provided a temporary harbour for the heart. Core family members keep in close contact with students regardless of how far away they live. Therefore, the tolerance and understanding of core family members could somewhat compensate for the uncomfortable social communication with the extended family. As mentioned by S07, "*They (members of the extended family) would ask 'are you still at school … why you haven't graduated?'. Things like that. I felt a little annoyed but not angry or frustrated … because my mom did so when she was at my age …* ". Thus, support from core family members helped resolve potential tensions in the students' social connections and benefited their mental wellbeing and social wellbeing.

Context

In this research, the context is complex, as it relates to multiple places in the constellation of "home". The participants moved among the old home, the (present) home and the alma mater when staying at home during the Spring Festival. Gatrell (2013) highlights that the context of mobility has a significant effect on wellbeing. We further explore the shift in the environment and context among multiple homes and the possible influence on student wellbeing. As shown in Figure 3, these places are located around or not far away from the (present) home in small cities, but the meanings that they carry and the people with whom the students meet vary. They provided the students different environments and contexts for mobility.

The significance of the "(present) home" for young university students can be summarized by Tuan's (1977) "intimate occasion". It is a place of nurture, where the needs of the students are heeded, their sorrow and confusion are understood, and their joy and happiness are naturally appreciated. As mentioned by S05, *"when I returned home, it's only my mom, dad, and younger brother there; they knew my situation and didn't ask awful questions about the job and boyfriend. So, I think it's okay, even though my aunts and uncles talked about it"*. Home provided the most private and intimate social environment, promoting the students' mental wellbeing, social wellbeing and self-continuity.

The old home is usually where grandparents or old relatives now live, and in most cases, the students had special childhood memories related to the "old home". As mentioned by S04, *"(One of the happiest things is) visiting my relatives in the village I lived in before. I don't mean that visiting my relatives itself is happy, but the chance to come back to that village, to take some photos in those old places … but visiting relatives is important because if they are not there, I'll never return to that village again"*. S02 mentioned the tradition of visiting the temple in the "old home": *"I'll go back and pray with my parents on 1 January; they believe it could bring good luck, and I just want to have a look at the temple. I used to go to the temple with grandma when I was young and lived in the 'old home', twice a month"*. Attachment to the old home is complex, including attachment to the relatives living there, to childhood memories, and to old traditions and beliefs. Both the landscape and familiar sights, sounds, and smells can trigger childhood memories, stimulate emotions of nostalgia and strengthen the sense of self-identity, benefiting the students' mental wellbeing and self-continuity.

The alma mater is where the students finished their primary or middle school education, and in general, it is in the town or city where the family currently resides. Visiting the alma mater positively influenced the students' wellbeing by improving their social wellbeing, mental wellbeing and self-development. Most of the time, visiting the alma mater was a group activity, a chance to reinforce social connections with previous classmates. In addition, compared with other locations, the alma mater was more likely to trigger a positive memory about the past. It can be considered an important station in students' lives, where they fought to gain admission to a good university. S11 mentioned the experience of visiting his alma mater with classmates, *"The school didn't change … there are many memories about us … it makes us think of the happy high school years … it's 7 or 8 years ago and we (friends) are still together after all these years … and I really appreciate it"*. This feeling resonates with the "positive time perception" noted by Pearce (2012); positive memories about the past and gratitude for what had happened promoted mental wellbeing and self-fulfilment.

However, it is worth noting that students' connection with context is mobile. On the one hand, the students are in a transitional life stage, and their ideologies and emotions sway between those of adolescents and adults. Home and familiar places might trigger happy memories and "positive time perception" but could also arouse self-reflection and

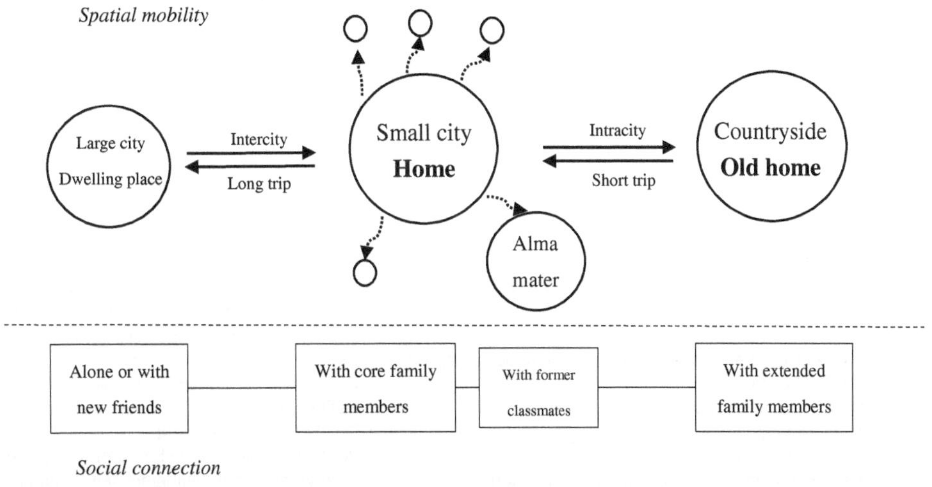

Figure 3. Relationships of different contexts in the Spring Festival homecoming.

life pressure. This complexity is reflected in S14's experience *"you'll see your playmates in childhood. Some of them may not live a good life, and you start to think everyone has different life courses. You might sigh for it. Life is different from the younger times. When you grow older, things become complicated"*. Although being at home and in familiar places is generally positively connected to wellbeing, there are moments when the student might feel stressed and uncertain about the future. On the other hands, society is also in transition due to high mobility. The connection between students and their "homes", especially their old homes, may change because of their mobile lifestyle. S12 mentioned that she does not have a hometown *"Once I went back home (old home in Hubei Province) with my parents (who usually live in Guangdong province). I threw up on New Year's Eve and spent 6 or 7 hours in the hospital. The doctor told me that it's because I was unaccustomed to the climate. That was the moment I felt that I don't belong to this place"*. When the connection to the "hometown" is weakened because of a mobile lifestyle, returning home may not provide positive wellbeing but only a sense of obligation.

Discussion

This paper has explored three dimensions of student mobility during the Spring Festival homecoming, including physical movement, social connections and contextual shift as well as their impacts on wellbeing. The slow life pace and changed body clock, tensions caused by working at home, and activities conducted both at home and beyond the home were analysed with regard to "physical movement". For "social connection", the people with whom the students met and socialized during the Spring Festival homecoming visit, their intensified social connections, and the pressure and tension caused by these intensified social connections were examined. Regarding the "context", the students interacted with multiple homes, that is, the (present) home, the old home and the alma mater, and they experienced mixed feelings of intimacy, nostalgia, a positive attitude towards life and complex feelings in a transitional stage.

The mobilities during the homecoming visit contributed to the students' wellbeing, both positively and negatively, in terms of physical wellbeing, mental wellbeing, social wellbeing, self-continuity and self-fulfilment (as shown in Table 4). Physical wellbeing is mainly influenced by the life pace at home and the activities engaged in during the visit. Here, wellbeing is connected not only to specific activities or body movements (Gatrell, 2013; Kou et al., 2017) but also to the life pace or body clock caused by the activities. Although potential tensions exist, in general, the students were able to smoothly adjust to the slow life pace at home or irregular arrangements during the Spring Festival and enjoyed their improved physical wellbeing. Mental wellbeing concerns a student's psychological perceptions and is related to all dimensions of mobility. Physical movement, social connections and context influence students' mental wellbeing.

Social wellbeing is closely connected to the people with whom the students met when moving or settling during the homecoming visit and their social interactions. Normally, the students receive support from their social relations at home, especially from core family members. However, tensions exist and may erupt during the Spring Festival because social connections are intensified during this special time period. Compared with students who are highly mobile and influenced by the modern lifestyle, the older generation has a more traditional attitude towards life. In their view, it is their responsibility to be concerned about the younger generation's romantic relationships, jobs, and marriage and to provide suggestions. However, students tend to desire privacy instead of inappropriate questions. In addition, this contradiction decreased the students' social wellbeing. The reasons underlying this contradiction are the generational gap and inappropriate intimacy rather than a lack of kinship and intimacy (Huang et al., 2017; Kou et al., 2017).

The influence of mobilities (physical movement, social connections and context) on self-continuity and

Table 4. Mobility dimensions, behavioural performance and wellbeing.

Dimension	Behaviour or viewpoint	Effects on wellbeing
Physical movement	Expectation regarding family reunion	Positive on mental wellbeing and self-continuity
	Long-distance and difficult trip	Negative on physical wellbeing
	Changed body clock	Negative on physical wellbeing
	Slow life pace	Positive on mental and physical wellbeing
	Double burn of work and leisure	Mixed on mental wellbeing, self-continuity and self-fulfilment
Social connection	Intensive social activities	Positive on mental and social wellbeing
	Pressure of personal affairs or privacy, home as controlling	Negative on social wellbeing
	Home as a harbour	Positive on mental and social wellbeing
Contextual shift	Mobility beyond daily activity space	Positive on mental and social wellbeing
	Home as family	Positive on all wellbeing
	Old home as community	Positive on mental wellbeing but negative on social wellbeing
	Alma mater as memory and spiritual motivation	Positive on social wellbeing, self-continuity and self-fulfilment

self-fulfilment reflect the students' feelings and experience during the transitional stage to adulthood. Self-continuity refers to the continuation of the previous lifestyle and a sense of familiarity. Visiting home, including the present home, the old home and the alma mater, provided the students an opportunity to experience the lifestyle that they used to enjoy, to reunite with friends and family members, and to recall childhood or school-time memories; these experiences promoted their self-continuity. However, they also needed to maintain a sense of continuity with regard to their lifestyle at college. They wanted to stay up late and declined joining some family activities to finish tasks from school, which can cause conflicts. Additionally, because they were accustomed to the modern lifestyle at school, they were concerned about their personal privacy, which constitutes a contradiction with the views of other family members. As noted in studies on emerging adulthood, the pursuit of freedom and privacy is a feature of the transitional stage (Arnett, 1994; Stephenson-Abetz & Holman, 2012). However, in Chinese culture, collective values are emphasized (Nelson et al., 2004); thus, students must find a balance between the self at home and the self at school, negotiate traditional collective values and the pursuit of privacy, and perform family duties while simultaneously completing their school work.

Self-fulfilment is mainly about pursuing self-development. Similar to the positive "time perception" mentioned by Pearce (2012), students may experience improved self-fulfilment by recalling positive memories and connecting them with the present and the future. However, one typical feature of the transitional stage is uncertainty about the future. Students normally do not have specific plans regarding a job, marriage or other issues that relatives at home are eager to know and give suggestions about. In addition, certain people or settings in the hometown might trigger self-reflection and strengthen a sense of uncertainty. Both relatives' expectations and students' uncertainty may decrease their self-fulfilment.

Conclusion

This paper explores the relationship between mobility and wellbeing from a behavioural perspective by focusing on the student mobility of homecoming during the Spring Festival. As a snapshot of contemporary society, the Spring Festival is discussed in terms of the coexistence of a family reunion and ideational conflict. This paper shows that students are eager to pursue an ideal home for their wellbeing, but they do not always obtain positive wellbeing from the homecoming. In fact, they suffer both physical and mental tensions. The Spring Festival homecoming means that students face a different rhythm of activity in their hometown, intensified social activities, and even unhappiness from paternalistic conversations. The mobility during the Spring Festival homecoming does not seem to be as happy and ideal as each person expects.

Theoretically, this study contributes to previous studies in three ways. First, college students' homecoming visits during the Spring Festival share a certain common ground with VFR or VHFP travel in terms of leaving one's dwelling place and visiting familiar people/places. Additionally, social relationships, the meaning of places, affirmations and obligations are important elements in both VFR and VHFP and in college students' homecoming visits. Thus, the mobility approach adopted in this study could also be used in VFR or VHFP studies to investigate visitors' experiences and wellbeing during trips. Linking visitors' mobility with their wellbeing answers the call to interpret the social meanings underpinning VFR travel. However, college students' homecoming visits also differ from general VFR or VHFP travel because students are in a transitional stage. Although previous VFR studies have noted some potential tensions between hosts and guests (e.g. Huang et al., 2017; Seaton, 2017), the tensions faced by college students during the Spring Festival homecoming visit differ from those previously identified. The sources of tension in college students' homecoming visits are related to the contradictions between the different lifestyles of the generations, excessive concerns by the hosts about the guests' future, and the guests' confusion and uncertainty in a transitional stage of life. These differences can enrich our understanding of VFR and VHFP travel.

Second, taking the Spring Festival homecoming visit as the specific example, this paper provides a snapshot of the happiness, gains, confusion, uncertainty and tensions faced by Chinese college students in a transitional life stage. They live away from home but are still connected to home or even multiple homes. Home provides them both joy and conflict. By adopting a mobility approach, this paper provides some insights into the sources of both this joy and these conflicts. To gain wellbeing in this transitional stage, students need to negotiate the past, the present and the future.

Third, this study offers insights into rethinking the meaning of home under rapid urbanization and mass rural-urban migration in China. In a society characterized by increased migration and high mobility, individuals often have more than one "home". Although the notion of "home" has been investigated in migrant studies (Kochan, 2016; Liu, 2014; Moskal, 2015), this paper focused specifically on multiple homes as a place of origin. By tracing the process of relocation, this paper identified three places that students felt associated with and attached to, that is, the (present) home, the old home and the alma mater. This identification may provide implications for studies that intend to investigate migrants' associations with their places of origin, for example, rural-urban floating labourers and highly educated migrants in China.

Practically, the mobility pattern and wellbeing involved in homecoming visits help us to understand the behaviour of VFR/diaspora tourism. The findings suggest a need for marketers to consider familism values and social connections when marketing VFR/diaspora tourism products. The rural areas where the old home or alma mater are located may be potential "destinations" during the festival (Moufakkir, 2011). Emotional connections to not only people but also places should be considered when marketing such destinations. On the other hand, the negative tensions identified from homecoming in this paper may provide insights for policymakers when encouraging domestic tourism or diaspora tourism. Policymakers should recognize the gap between returners and local people and use a strategy to encourage positive integration. Promoting wellbeing from homecoming mobilities is the ultimate goal for students, marketers and policymakers (Li & Chan, 2017).

Methodologically, this paper is intended to demonstrate the usefulness of behavioural methods for exploring both subjective and objective behaviours in the long term and on a daily basis. GPS tracking can provide clear behavioural information but ignore subjective information, which is crucial in the social sciences (Frändberg, 2008, 2014; Shoval et al., 2018). This study combines a quantitative behavioural method and qualitative interviews through coordination via the theoretical paradigm of time geography. Therefore, it applies a mixed method that is not strictly quantitative or qualitative. Due to the variable scale of the space-time path, this study describes long-term behaviour between the dwelling city and the home city and constructs a space-time path on a daily basis to explore the day-to-day variability and spatial scale.

This study also has some limitations. The sample size was relatively small, even though a two-stage survey was conducted. It tried to select a sample of students that varied by place of birth and hometown location, but it ignored the similar educational backgrounds and processes. Further quantitative and qualitative studies should be conducted in different universities or with students with different majors. In addition, this study was carried out in the Chinese Spring Festival context. Whether the conclusions can remain valid in other cultural contexts should also be examined.

Disclosure statement

No potential conflict of interest was reported by the authors.

Funding

This research received grants from the National Science Foundation of China (No. 41501172; 41771145).

ORCID

Honggang Xu http://orcid.org/0000-0003-1354-6654

References

Arnett, J. J. (1994). Are college students adults? Their conceptions of the transition to adulthood. *Journal of Adult Development*, *1*(4), 213–224.

Arnett, J. J. (2007). Emerging adulthood: What is it, and what is it good for? *Child Development Perspectives*, *1*(2), 68–73.

Braun, V., & Clarke, V. (2006). Using thematic analysis in psychology. *Qualitative Research in Psychology*, *3*(2), 77–101.

Cairns, D. (2008). Moving in transition: Northern Ireland youth and geographical mobility. *Young*, *16*(3), 227–249.

Chan, K. W., Liu, T., & Yang, Y. (1999). Hukou and non-hukou migrations in China: Comparisons and contrasts. *International Journal of Population Geography*, *5*(6), 425–448.

Cui, C., Geertman, S., & Hooimeijer, P. (2015). Residential mobility of skilled migrants in Nanjing, China. *Environment and Planning A*, *47*(3), 625–642.

Duyvendak, W. (2011). *The politics of home*. Basingstoke: Palgrave.

Fang, C., Yang, D., & Meiyan, W. (2009). Migration and labor mobility in China. *Human Development Research Paper (HDRP) Series*, *09*, 1–38.

Frändberg, L. (2008). Paths in transnational time-space: Representing mobility biographies of young Swedes. *Geografiska Annaler: Series B, Human Geography*, *90*(1), 17–28.

Frändberg, L. (2014). Temporary transnational youth migration and its mobility links. *Mobilities*, *9*(1), 146–164.

Freedman, V. A., Carr, D., Cornman, J. C., & Lucas, R. E. (2017). Aging, mobility impairments and subjective wellbeing. *Disability and Health Journal*, *10*(4), 525–531.

Fuligni, A. J. (2007). Family obligation, college enrollment, and emerging adulthood in Asian and Latin American families. *Child Development Perspectives*, *1*(2), 96–100.

Gatrell, A. C. (2011). *Mobilities and health*. Aldershot, UK: Ashate.

Gatrell, A. C. (2013). Therapeutic mobilities: Walking and 'steps' to wellbeing and health. *Health & Place*, *22*, 98–106.

Griffin, T. (2013). Research note: A content analysis of articles on visiting friends and relatives tourism, 1990–2010. *Journal of Hospitality Marketing & Management*, *22*(7), 781–802.

Hägerstrand, T. (1970). What about people in regional science? *Papers in Regional Science*, *24*(1), 7–24.

He, S. (2014). Consuming urban living in 'villages in the city': Studentification in Guangzhou, China. *Urban Studies*, *52*(15), 2849–2873.

Huang, W. J., King, B., & Suntikul, W. (2017). VFR tourism and the tourist gaze: Overseas migrant perceptions of home. *International Journal of Tourism Research*, *19*(4), 421–434.

Janta, H., Cohen, S. A., & Williams, A. M. (2015). Rethinking visiting friends and relatives mobilities. *Population, Space and Place*, *21*(7), 585–598.

Jones, T., Harms, L., & Heinen, E. (2016). Motives, perceptions and experiences of electric bicycle owners and implications for health, wellbeing and mobility. *Journal of Transport Geography*, *53*, 41–49.

Kahneman, D., Diener, E., & Schwarz, N. (Eds.). (1999). *Well-being: Foundations of hedonic psychology*. New York, NY: Russell Sage Foundation.

Kochan, D. (2016). Home is where I lay down my hat? The complexities and functions of home for internal migrants in contemporary China. *Geoforum*, *71*, 21–32.

Kou, L., Xu, H., & Hannam, K. (2017). Understanding seasonal mobilities, health and wellbeing to Sanya, China. *Social Science & Medicine*, *177*, 87–99.

Kou, L., Xu, H., & Kwan, M.-P. (2018). Seasonal mobility and well-being of older people: The case of 'Snowbirds' to Sanya, China. *Health & Place*, *54*, 155–163.

Kwan, M. P., & Schwanen, T. (2016). Geographies of mobility. *Annals of the American Association of Geographers*, *106*(2), 243–256.

Li, M. (2014). Ritual and social change: Chinese rural–Urban migrant workers' Spring Festival homecoming as secular pilgrimage. *Journal of Intercultural Communication Research*, *43* (2), 113–133.

Li, T., & Chan, T. (2017). Diaspora tourism and well-being: A eudaimonic view. *Annals of Tourism Research*, *63*, 205–206.

Liu, L. S. (2014). A search for a place to call home: Negotiation of home, identity and senses of belonging among new migrants from the People's Republic of China (PRC) to New Zealand. *Emotion, Space and Society*, *10*, 18–26.

Markowitz, F., & Stefansson, A. H. (Eds.). (2004). *Homecomings: Unsettling paths of return*. Lanham, MD: Lexington Books.

Moskal, M. (2015). 'When I think home I think family here and there': Translocal and social ideas of home in narratives of migrant children and young people. *Geoforum*, *58*, 143–152.

Moufakkir, O. (2011). Diaspora tourism: Using a mixed-mode survey design to document tourism behavior and constraints of people of Turkish extraction resident in Germany. *Journal of Vacation Marketing*, *17*(3), 209–223.

Nelson, L., Badger, S., & Wu, B. (2004). The influence of culture in emerging adulthood: Perspectives of Chinese college students. *International Journal of Behavioral Development*, *28*(1), 26–36.

Pearce, P. L. (2012). The experience of visiting home and familiar places. *Annals of Tourism Research*, *39*(2), 1024–1047.

Ryan, R. M., & Deci, E. L. (2001). On happiness and human potentials: A review of research on hedonic and eudaimonic well-being. *Annual Review of Psychology*, *52*(1), 141–166.

Seaton, T. (2017). Qualitative approaches to the phenomenology of VFR travel: The use of literary and cultural texts as resources. *International Journal of Tourism Research*, *19*(4), 455–465.

Seeborg, M. C., Jin, Z., & Zhu, Y. (2000). The new rural-urban labor mobility in China: Causes and implications. *The Journal of Socio-Economics*, *29*(1), 39–56.

Seiter, L. N., & Nelson, L. J. (2011). An examination of emerging adulthood in college students and nonstudents in India. *Journal of Adolescent Research*, *26*(4), 506–536.

Shen, Y., Kwan, M. P., & Chai, Y. (2013). Investigating commuting flexibility with GPS data and 3D geovisualization: A case study of Beijing, China. *Journal of Transport Geography*, *32*, 1–11.

Shoval, N., Schvimer, Y., & Tamir, M. (2018). Real-time measurement of tourists' objective and subjective emotions in time and space. *Journal of Travel Research*, *57*(1), 3–16.

Stanley, J. K., Hensher, D. A., Stanley, J. R., & Vella-Brodrick, D. (2011). Mobility, social exclusion and well-being: Exploring the links. *Transportation Research Part A: Policy and Practice*, *45*(8), 789–801.

Stefansson, A. H. (2004). Homecomings to the future: From diasporic mythographies to social projects of return. In F. Markowitz & A. H. Stefansson (Eds.), *Homecomings: Unsettling paths of return* (pp. 2–20). Lanham, MD: Lexington Books.

Stephenson-Abetz, J., & Holman, A. (2012). Home is where the heart is: Facebook and the negotiation of "old" and "new" during the transition to college. *Western Journal of Communication*, *76*(2), 175–193.

Tuan, Y. F. (1977). *Space and place: The perspective of experience*. London, UK: University of Minnesota Press.

Tuan, Y. F. (2007). *Coming home to China*. London, UK: University of Minnesota Press.

Urry, J. (2007). *Mobilities*. Cambridge, UK: Polity Press.

Waterman, A. S. (1993). Two conceptions of happiness: Contrasts of personal expressiveness (eudaimonia) and hedonic enjoyment. *Journal of Personality and Social Psychology*, *64*(4), 678–691.

Zhao, Y., Dijst, M., & Chai, Y. (2016). Between haven and heaven in cities: A comparison between Beijing (China) and Utrecht (the Netherlands). *Urban Studies*, *53*(12), 2469–2487.

Ziegler, F., & Schwanen, T. (2011). 'I like to go out to be energised by different people': An exploratory analysis of mobility and wellbeing in later life. *Ageing & Society*, *31*(5), 758–781.

Modeling the perception of walking environmental quality in a traffic-free tourist destination

Linchuan Yang, Xu Wang, Guibo Sun and Yuan Li

ABSTRACT

This study aims to determine how tourists perceive the quality of a walking system in a traffic-free destination. We conducted a questionnaire survey on tourist satisfaction with attributes and overall performance of the walking system at Kulangsu (China) in October 2017 and February 2018. We developed an ordered probit model to evaluate the relative importance of each attribute by employing the pooled cross-sectional data. Results show that the attributes that need an enhancement in the view of tourists are the level of crowding and cleanliness. The two attributes have not been pointed out in previous literature. Accordingly, policy implications are proposed. The findings enhance our understanding of tourist satisfaction with the walking system.

Introduction

Tourist satisfaction with a destination is essential to consumption-related fulfillment and successful destination marketing (Huh, Uysal, & McCleary, 2006; Oliver, 2014) and thus plays a crucial role in the attractiveness, vitality, and profitability of a destination (Anderson, Fornell, & Lehmann, 1994). Understandably, it is affected by a great many attributes offered at the destination level (e.g. tour leaders, shopping activities, travel experience, variety of activities offered, and communication through native language) (Hall, O'Mahony, & Gayler, 2017; Huh et al., 2006; Parahoo, Harvey, & Radi, 2014). These attributes have the potential to motivate or pull the tourist towards one destination over another (Hall et al., 2017). Additionally, intra-destination travel system, which affects mobility, the ability to move from one place to another independently and safely (Shoval & Isaacson, 2007), is a crucial tangible attribute of a destination. Whether or not the travel system meets tourists' mobility needs affects travel experience and tourist satisfaction.

Traffic-free (or pedestrian-only) destination (e.g. the Great Wall, the Forbidden City, and Golden Temple), which is only open to pedestrians, is a subset of tourist destinations. The travel (or walking) system of this category of destinations effectively connects all scenic spots, thereby simultaneously ensuring "travel for tourism" and "travel as tourism." In the context, "the journey itself, as much as the destination, may be part or even the prime attraction of the trip", and "tourism is transport" (Speakman, 2005, p. 129). Thus, tourist satisfaction with (or perception of) the walking environmental quality (or pedestrian satisfaction) in this subset should be well understood. However, there are few studies on pedestrian satisfaction in traffic-free tourist destinations in existing literature, as far as we are aware.

In light of this, we investigated walking environmental quality satisfaction (a specific type of tourist satisfaction) and modeled the relationship between service quality attribute satisfaction and overall satisfaction. We aim to address the following research questions:

- Which attribute has the largest impact on tourists' satisfaction level with the walking system?
- Which attributes need to be improved?
- How do we improve these attributes?

We conducted an interviewer-administered face-to-face survey in October 2017 and February 2018 in Kulangsu (a UNESCO (United Nations Educational, Scientific and

Cultural Organization) World Heritage Site in China, also known as Gulangyu and Gulang Island). We aimed to obtain the tourists' levels of satisfaction with individual service quality attributes and overall performance of the walking system. Using the data, we developed an ordered probit model (OPM) to estimate the importance of each service quality attribute to the overall service quality. We validated the OPM specification by performing a first preference recovery (FPR) analysis to gain confidence in the model performance and ensure prediction accuracy (Ben-Akiva & Lerman, 1985). As satisfaction with a service is based on service quality which is thought to be within the control of service suppliers, we proposed policy implications based on the modeling result.

This study contributes to the existing literature on walking satisfaction and walkability of (traffic-free) tourist destinations in the following aspects: (1) contributing to the limited research on pedestrian satisfaction in a traffic-free tourist destination; (2) identifying the walking attributes that substantially influence tourist satisfaction with the overall performance; (3) presenting an analytical framework to investigate barriers that tourists encounter; (4) determining priorities for actions to improve the quality of the walking environment and maximize the new investments effects on tourist satisfaction; and (5) suggesting policy measures to strengthen pedestrian planning in Kulangsu and other destinations with a similar travel system.

The remainder of the paper proceeds as follows. The ensuing section outlines the related literature on (1) attribute satisfaction and overall satisfaction of tourists and (2) perceived built environment correlates of walking. "Data" section introduces the study area and the data collection process. It also tabulates the participants' profile and satisfaction levels. "Methodology" section outlines OPM and FPR analysis. "Results" section presents the corresponding results. "Conclusions and implications" section concludes the paper, discusses insights for policy intervention, and suggests research strengths and limitations.

Literature review

Attribute satisfaction and overall satisfaction of tourists

Tourist satisfaction typically refers to the satisfaction of tourists with a destination. It is often considered as the outcome of an experiential quality of a trip (Spreng, MacKenzie, & Olshavsky, 1996) or the "emotional state after exposure to the opportunity" (Baker & Crompton, 2000, p. 785)" and is strongly correlated with the long-term success of tourist destinations (Masiero, Qian, Fong, & Law, 2018). It is a function of various attributes offered by the destination. A plethora of literature has focused on the link between attribute satisfaction and overall satisfaction of tourists and identified a multitude of determinants of overall satisfaction. Most studies, however, are conducted in cities of North America and Europe. Understandably, tourist satisfaction may be affected by contexts and socio-economic and demographic characteristics of tourists. This leads to a lack of generalizability of findings in Western cities to Asian cities.

Utilizing structural equation modeling, Spreng et al. (1996) contended that attribute satisfaction is a direct antecedent of overall satisfaction. Based on 310 questionnaire surveys collected from Australians who visited Vietnam, Truong and Foster (2006) maintained that overall satisfaction is related to a wide array of destination attributes, such as local transport and climate. Huh et al. (2006) utilized a sample of 201 tourists visiting the Virginia Historic Triangle (the USA) and confirmed that overall satisfaction is influenced by satisfaction with a series of attributes, such as cultural/heritage attraction, general tour attraction, shopping attraction, and information. Hui et al. (2007) identified that significant contributory attributes (e.g. accommodation and food) for to-Singapore tourists coming from different regions, namely Asia, North America, Europe, and Oceania. They further concluded that different tourists value the attributes differently. Based on a sample of 500 Chinese tourists visiting the Republic of Korea and path analysis, Lee et al. (2011) demonstrated that tourist satisfaction is affected by a host of destination attributes, including scenery, historical and cultural legacy, comfort and safety, and cost-effectiveness. Song, Van der Veen, Li, and Chen (2012) employed structural equation modeling and proposed that attractions, hotels, immigration, restaurants, retail shops, and transportation significantly influence tourist satisfaction. Hall et al. (2017) used a sample of 426 tourists who visited ski resorts in Australia and employed factor analysis and structural equation modeling to confirm the relation between attribute satisfaction and overall satisfaction. Contributory attributes include transport, employees, slopes, fun/action, and accommodation. Masiero et al. (2018) used a sample of 7431 tourists to Macao, developed a set of OPMs to determine correlates of overall satisfaction and observed that uniqueness of destination, attractiveness of casinos, information availability, ease of transport, short vacation, and value for budget are significantly related to overall satisfaction.

A few studies focus on a specific type of tourist satisfaction (e.g. satisfaction with shopping at a destination) rather than tourist satisfaction with a destination as a whole. Bowie and Chang (2005) found that tourist satisfaction with a guided package tour is influenced by the role of the tour leader and the service performance. Tosun, Temizkan, Timothy, and Fyall (2007) identified the determinants of tourist satisfaction with shopping (e.g. variety of products, authenticity, and staffs' language ability). Murphy, Moscardo, Benckendorff, and Pearce (2011) intimated that tourist satisfaction with the retail experience in a typical tourist shopping village is affected by store, service, and product. Oviedo-García, Vega-Vázquez, Castellanos-Verdugo, and Reyes-Guizar (2016) suggested that satisfaction with souvenir shopping is influenced by internal attraction, service differentiation, service provision, and external attraction. However, to the best of our knowledge, there are no studies devoted to investigating the determinants of tourist satisfaction with the intra-destination travel system in a traffic-free destination.

Perceived built environment correlates of walking

Walking is a moderate physical activity that has a wide array of known health, environmental, economic, and social benefits (Leyden, 2003; Manson et al., 1999; Xu et al., 2016; Guo et al., 2017; Lu et al., 2018; Yang et al., 2018). Recently, there has been mounting research on built environment correlates of walking (Saelens & Handy, 2008). Documented evidence, mostly from transportation planning and public health fields, has shown that the built environment could either facilitate or hinder walking behavior (Saelens & Handy, 2008).

We review published scholarly work on "perceived" built environment determinants of walking. Note that "perceived" built environment determinants are significantly associated with walking at slightly higher rates than "objective" environmental factors (Orstad, McDonough, Stapleton, Altincekic, & Troped, 2017). More complete reviews on (both perceived and objectively measured) built environment correlates of walking can be found in Saelens and Handy (2008), Brownson, Hoehner, Day, Forsyth, and Sallis (2009), and Orstad et al. (2017).

Table 1 is the review of the selected studies. A quick-reference format is utilized for ease of use by interested readers. About data acquisition methods, these studies highly relied on self-report survey instruments. In this sense, they are dramatically different from their counterparts on objective environmental factors, which usually collect data using observational techniques, such as street audits and Geographic Information System. In addition, most were cross-sectional studies and very few adopted longitudinal (or panel) research designs.

A wide variety of built environment correlates of walking has been determined, including but not limited to density, safety, connectivity, comfort, aesthetics, functionality, the presence of benches, readability of signage, and land use diversity. However, the findings are usually inconsistent. Even conflicting outcomes are found. Note that the majority of existing literature is empirically conducted in advanced economies, such as the USA, Europe, and Australia. These contexts often have low or medium levels of population density settings (e.g. the USA and Australia). Understandably, walking conditions/behavior/preferences in high-density places (e.g. Tokyo, Singapore, Seoul, and virtually all Chinese cities) are considerably distinguished from those in low- and medium-density counterparts. These contexts have received much less scholarly attention. Moreover, existing literature usually investigates walking perception of either all city residents (e.g. Cerin, Leslie, Du Toit, Owen, & Frank, 2007) or a specific group (e.g. African-American by Bopp et al., 2006). Comparatively, the segment of tourists has rarely been studied. Most strikingly, most of the related studies, especially those on pedestrian satisfaction (e.g. Choi, Kim, Min, Lee, & Kim, 2016; Jensen, 2007; Kim, Choi, & Kim, 2013; Kim, Park, & Lee, 2014; Wang, Li, Wang, & Namgung, 2011) have been conducted in multimodal contexts, in which other travel modes (e.g. automobile) are readily available. Since walking is the only travel mode in traffic-free places, the insights gained from the extant literature that focuses on facilitating walking behavior and decreasing motorized travel cannot be generalized.

Data

Study area

Kulangsu is a famous traffic-free tourist destination in China with outstanding colonial landscape and various cultural attractions. It is a tiny island situated at the east of the coast of Xiamen (Figure 1), a sub-provincial city located on the southeast coast of China. Xiamen comprises six districts and owns a permanent population of 3.92 million and an administrative area of 1699.39 km^2 as of the end of 2016. Enticingly called the "Garden on the Sea", it is a famous livable, comfortable, and sustainable city in China (Xu and Yang, 2019). Similar to a vast sea of cities in developing countries, an important industry of Xiamen is tourism, mainly due to the intersection of historical and contemporary cultures and stunning natural and cultural landscapes.

Table 1. Review of perceived built environment correlates of walking.

Reference	Sample	Data acquisition method	Environmental factor(s) concerned	Findings
Bopp et al. (2006)	572 African-American in a congregation in South Carolina, the USA	Interviewer-administrated telephone survey	Walkability, crime, availability of sidewalks and street lighting, and the presence of public parks	Environmental factors are not significant to affect walking activity.
Burton, Turrell, Oldenburg, and Sallis (2005)	1827 people aged between 18 and 65 in Brisbane, Australia	Self-administered mail survey	Physical attributes (e.g. public transport and streetlights), aesthetic features (e.g. cleanliness), traffic (e.g. busyness of streets), and facilities (e.g. gyms and pools)	Environmental factors explain 0.6% of the variability in walking activity.
Cao, Handy, and Mokhtarian (2006)	1368 respondents in Austin, the USA	Self-administered mail survey	Safety, shade, scenery, traffic, local store quality, walking advantage, and walking comfort	Safety, shade, and traffic are related to strolling frequency.
Cerin et al. (2007)	2194 participants in Adelaide, Australia	Self-administered mail survey	Proximity of 24 types of destinations (e.g. workplace and school)	Proximity to commercial destinations and workplaces significantly affect transportation walking activity.
De Bourdeaudhuij, Sallis, and Saelens (2003)	521 participants in Ghent, Belgium	Self-administered mail survey	Residential density, land use diversity, access to local shopping and transit, availability of sidewalks and bike lanes, aesthetics, safety from crime and traffic, connectivity, and neighborhood services	Availability of sidewalks affects men's walking behavior while access to shops, facilities, and public transit affect women's walking behavior.
De Bourdeaudhuij, Teixeira, Cardon, and Deforche (2005)	247 people in Oeiras, Portugal, and 279 people in Ghent, Belgium	Self-administrated off-site questionnaire survey	Same as above	Availability of sidewalks (for Portuguese sample) and land use diversity (for Belgian sample) play a role in determining recreational walking.
Hoehner, Ramirez, Elliott, Handy, and Brownson (2005)	1053 people in two regions (St. Louis and Savannah) of the USA	Interviewer-administrated telephone survey	Accessibility to opportunities, recreational facility availability, transport environment (e.g. the presence of sidewalks and bike lane), and neighborhood aesthetics (e.g. shade and cleanliness).	Transportation physical activity is positively related to access to opportunities and presence to bike lanes but negatively related to neighborhood aesthetics. Recreational physical activity is positively related to perceived availabilities of recreational facility.
Reed, Wilson, Ainsworth, Bowles, and Mixon (2006)	1148 people in a rural community of Sumter County, the USA	Interviewer-administrated telephone survey	The presence of sidewalks	The presence of sidewalks is related to physical activity patterns of whites, but this is not the case for non-whites.
Spence et al. (2006)	3144 Canadians visiting the website of *Canada on the Move*	Internet-based survey	Access to shops and transit, the presence of sidewalks, crime, recreation facility availability, scenery, intersection, and walking accessibility to opportunities	Interesting scenery and decent walking accessibility to opportunities are associated with high likelihood of sufficient walking for women. However, this outcome cannot apply to men.
Sun, Han, Sun, and Oreskovic (2018)	805 children lived within the school catchment neighborhoods in Shenzhen, China	Interviewer-administrated face-to-face survey	Permeability, route across public open space, route across points of interest, aesthetics, and safety	All factors except permeability exert significant influences on active commuting mode choice of children.
Suminski, Poston, Petosa, Stevens, and Katzenmoyer (2005)	474 residents of a large Midwestern metropolitan Area of the USA	Door-to-door interview	Functionality (e.g. the integrity of sidewalks), safety (from crime and traffic), aesthetics (i.e. cleanliness and building view), and accessibility to destinations	For women, safety significantly affects both exercise and with-dog walking while accessibility to destinations affects transportation walking. For men, only one factor, namely aesthetics, affects transportation walking.
Voorhees et al. (2010)	890 eighth-grade girls of the USA	Interviewer-administrated telephone survey	Safety, aesthetics, and access to facilities near home	Safety and access to facilities near home are associated with the walking activity.

Figure 1. Location of Kulangsu.

Kulangsu has a total area of 1.87 km^2 and accommodates approximately 14 thousand residents (Li et al., 2016, 2017). This island was formed as an international settlement in 1903 and has become an important channel for Sino-foreign exchanges since then. It has a naturally generated spatial structure and a blend of eastern and western styles. This reflects a long historical development process with outstanding cultural diversity. Moreover, it has a nickname of "the Island of Music" and is thus often regarded as a haven for music-themed travels. As a popular domestic tourist destination, Kulangsu serves 25,000 to 65,000 tourists each day. Bestowed with rich cultural and historical heritage, Kulangsu was recognized as a National Scenic Area by the State Council of China in 1988. Further, it was described as the most picturesque district by the Chinese National Geography magazine and was given "5A" National Tourist Attraction status by the Chinese National Tourism Administration in 2007. Noticeably, in July 2017, the island was successfully listed as a UNESCO World Heritage Site on July 8th, 2017. Longtou Road Area, with numerous small catering businesses (an indicator of areal vibrancy, see Ye et al., 2018), is the heart of this island.

A distinguishing feature of Kulangsu is traffic-free (Figure 2): automobiles and bicycles are banned. Due to the strict vehicular restriction, walking almost becomes the exclusive mode of travel.

Mainly due to being listed as a World Heritage Site, tourist volume of Kulangsu has been substantially increased (Figure 3) and is predicted to increase in the upcoming years, while the local government has been actively promoting sustainable tourism development. Therefore, how tourists perceive the quality of intra-destination travel (walking) system and improve the system to better serve tourist sightseeing experience should be investigated.

Data collection

One pilot survey was conducted in September 2017 to select the service quality factors examined and refine our

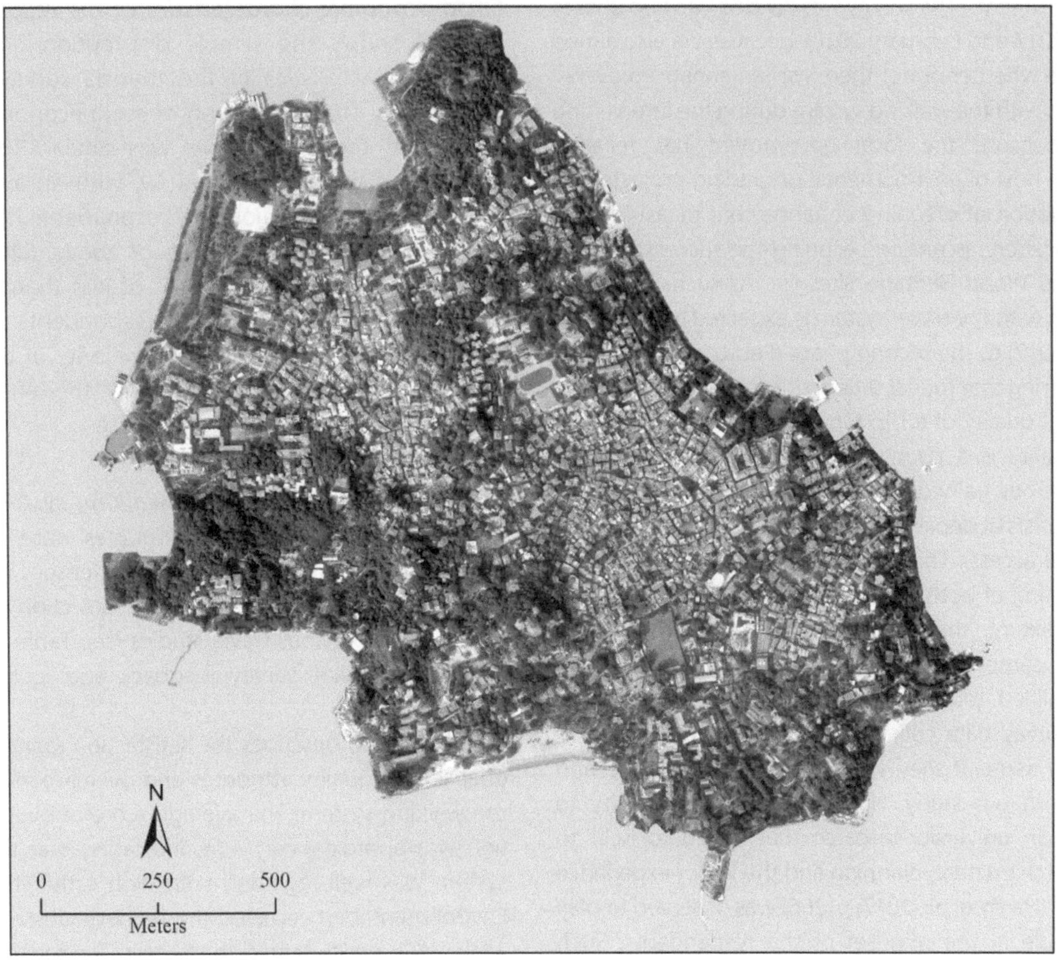

Figure 2. A satellite image of Kulangsu.

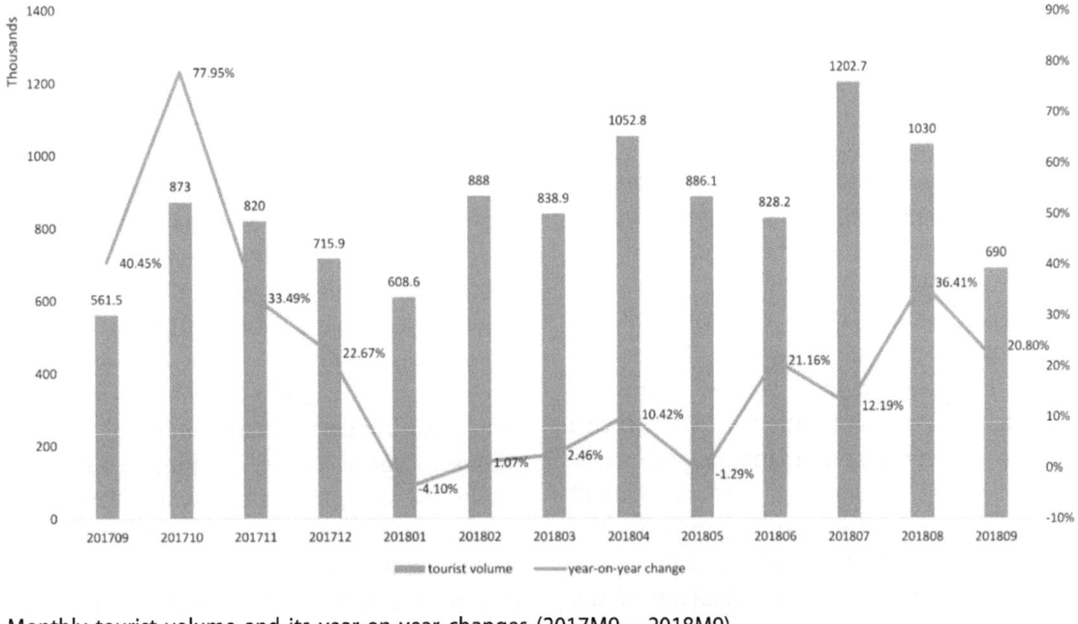

Figure 3. Monthly tourist volume and its year-on-year changes (2017M9 – 2018M9).

Note: made by the authors using the data from the official website of Kulangsu (http://www.gly.cn/index.php/cms/item-list-category-1374.shtml)

survey instruments. The survey was divided into two phases (October 2017 and February 2018), because we also aimed to examine whether or not there was an enhanced overall satisfaction with the walking system during the intervening months, because the local government has recently launched a host of environmental upgrading projects (e.g. implementation of greening enhancement measures and on-route scenery ascension), echoing the successful inclusion on the World Heritage Sites list. As such, enhanced satisfaction with the travel system is expected from the first (October 2017) to the second phase (February 2018).

Considering that tourist satisfaction is the outcome of an experiential quality of a trip (Spreng et al., 1996) and that service quality is a cognitive assessment of the served experience, our trained interviewers only randomly interviewed tourists at departure at two ferry docks (points with the highest access). This segment of tourists had a better understanding of both the island and the walking system. To some extent, the fact that there are only 2 docks decreases sample selection bias. Interviewers approached qualified tourists (those at departure), explained that the survey data collected are only for research purposes, and asked if they were willing to cooperate and participate in our study. Note that "the opportunity to participate in university research that aimed to help to improve infrastructure, planning and the visitor experience in general" (Hardy et al., 2017, p. 266) was observed to play a crucial role in the mindset of the respondents. Such mentality helped increase the quantity and improve the quality of their inputs. Notably, the active participation is due in part to the cultural background of Chinese respondents (Hardy et al., 2017). After finishing the survey, we offered a small souvenir to each respondent, which served as an incentive for participation.

The questionnaire items included: (1). Socio-economic characteristics of the tourists (e.g. gender, age, and education); (2). Attribute satisfaction (e.g. the level of crowding and cleanliness) and overall satisfaction with the walking system, both of which are measured using an ordinal scale from 1 (highly dissatisfied) to 5 (highly satisfied). Further, open questions regarding reasons for dissatisfaction with each attribute are also asked.

In total, 623 participants (311 in October 2017 and 312 in February 2018) finished the survey. The pooled cross-sectional data are fed into ordered regression modeling to identify the contribution of each service quality attribute to overall performance as well as to discern whether there is significant satisfaction uplift between the two phases spanning five months. The Cronbach's alpha value of the data is 0.76, which implies that internal consistency (or reliability) of the data is at the acceptable level.

Socio-economic characteristics of the respondents

Table 2 shows the sample distribution of socioeconomic characteristics of the tourists surveyed in the two phases. The distribution of socio-economic characteristics of the respondents was similar. The respondents aged between 16 and 69, with an average age of around 31. The majority of respondents had a junior college/undergraduate degree or above. 52% of participants had a personal income of less than 5000 RMB per month. Over 90% of the respondents performed independent tour style and the majority of the respondents took a day tour. Interestingly, the length of stay became shorter in the second phase.

Tourist satisfaction with the walking system

A total of 10 service quality attributes were selected to address the circumstances and characteristics of travel on the island. Notably, the attribute selection process is informed by previous travel studies (see Table 3) and also incorporates pilot survey feedback and local transport experts' advice.

Table 3 also describes the satisfaction ratings of individual service quality attributes and overall performance of the walking system. The average score of overall satisfaction was approximately 4.15, indicating that the walking system was well regarded with high satisfaction ratings. Environmental art received the highest satisfaction score, while WIFI service scored the lowest. The satisfaction level of the respondents was the most divergent regarding WIFI service, possibly due to their dissimilar preferences. Notably, a salient enhancement of the overall performance of the walking system can be observed.

Methodology

Model calibration: ordered probit model

The outcome variable in this study is overall satisfaction, measured using an ordinal scale ranging from 1 (highly dissatisfied) to 5 (highly satisfied). As a multifacet construct, overall satisfaction with a service is affected by satisfaction with all aspects of the service provided (Olawole & Aloba, 2014).

For analysis of data with the inherent ordering property, we cannot adopt ordinary least squares (OLS) regression models which minimize the sum of the square of error terms, as they would yield biased results. As such, in this study, an OPM was calibrated to evaluate the relative importance of a host of service quality attributes to the overall service performance. This approach dates from Aitchison and Silvey (1957) and has been extensively applied in a number of cases

Table 2. Participants profiles.

Personal particular	First phase (N = 311)		Second phase (N = 312)	
	Frequency	Percentage	Frequency	Percentage
Gender				
Male	122	39.23%	148	47.44%
Female	189	60.77%	164	52.56%
Education				
Junior high school or below	17	5.47%	40	12.82%
Senior high school	69	22.19%	106	33.97%
Junior college/undergraduate	215	69.13%	162	51.92%
Postgraduate or above	10	3.22%	4	1.28%
Tour style				
Independent tour	283	91.00%	291	93.27%
Package tour	28	9.00%	21	6.73%
Household car availability				
Yes	234	75.24%	241	77.24%
No	77	24.76%	71	22.76%
Personal monthly income (RMB)				
0–3000	82	26.37%	63	20.19%
3001–5000	81	26.05%	101	32.37%
5001–7000	67	21.54%	83	26.60%
7001+	81	26.04%	65	20.83%
	Mean	Std. Dev.	Mean	Std. Dev.
Age	32.31	10.50	29.24	6.33
Length of stay (days)	1.47	1.14	1.21	0.59

Table 3. Tourist satisfaction ratings on individual service quality attributes and overall performance of the walking system.

Service quality attribute	References	First phase (N = 311)		Second phase (N = 312)	
		Mean	Std. Dev.	Mean	Std. Dev.
Convenience (e.g. few obstacles)	De Bourdeaudhuij et al. (2003), 2005)	4.25	0.70	4.13	0.65
Comfort	Litman (2003); Cao et al. (2006)	4.24	0.68	4.12	0.67
The presence of shade (e.g. trees and arcade)	Hoehner et al. (2005); Cao et al. (2006); Tian and Ewing (2017)	4.25	0.74	4.24	0.70
Street lighting	Burton et al. (2005); Bopp et al. (2006); Tian and Ewing (2017)	4.11	0.72	4.09	0.68
Pavement	Reed et al. (2006); Spence et al. (2006)	4.27	0.69	4.14	0.71
Cleanliness	Burton et al. (2005); Hoehner et al. (2005); Suminski et al. (2005)	4.31	0.66	4.24	0.59
The level of crowding	Li, Yang, Shen, and Wu (2019)	4.14	0.81	4.14	0.68
Signage	Li et al. (2019)	4.14	0.84	4.16	0.70
Environmental art	Cervero and Duncan (2003); Orstad et al. (2017)	4.25	0.75	4.31	0.66
WIFI service	Wang and Loo (2018)	3.83	0.97	4.20	0.70
Overall performance	-	4.05	0.56	4.25	0.60

with an ordinal outcome variable (e.g. Masiero et al., 2018).

The OPM assumes the existence of a latent variable y^*, which depends linearly on explanatory variables. Their relationship can be written as follows:

$$y^* = \sum_m \beta_m x_m + \varepsilon,$$

where x_m is the score of quality attribute m, β_m is the corresponding coefficient, and ε is a residual, which follows the standard normal distribution and is independent from β_m . In our case, theoretically, all coefficients are non-negative since individual service quality should have positive or negligible repercussion on the overall service quality.

The observed ordinal outcome variable y (the score of the walking system in our case) is in category j when y^* falls in the jth interval. Their relationship can be summarized as follows:

$y = j$ if $a_{j-1} < y^* \le a_j$,

where a_j is the jth intercept $(-\infty = a_0 < a_1 < \ldots < a_J = \infty)$.

Based on these assumptions, the probability that the observation chooses alternative j, $P(y = j)$, can be written as:

$$P(y = j) = P(a_{j-1}\, y^* \le a_j) = P(a_{j-1} \sum_m \beta_m x_m + \varepsilon \le a_j)$$

$$= P(a_{j-1} - \sum_m \beta_m x_m < \varepsilon \le a_j - \sum_m \beta_m x_m)$$

$$= \int_{a_{j-1}-\sum_m \beta_m x_m}^{a_j-\sum_m \beta_m x_m} f(k)dk$$

$$= F(a_j - \sum_m \beta_m x_m) - F(a_{j-1} - \sum_m \beta_m x_m),$$

where $f(.)$ and$F(.)$ are density function of and cumulative distribution function, respectively, of the standard normal distribution. The coefficients and intercepts are simultaneously estimated by maximizing the log-likelihood function.

Model validation: first preference recovery analysis

Validation is part and parcel of the model development process for gaining confidence in the model performance, but it is usually ignored due to lack of monetary and time resources (Ben-Akiva & Lerman, 1985).

FPR analysis takes a "hold-out" approach: an estimation on a subset of the data which is then used to "predict" the responses for another dataset (i.e. the withheld subset of the data). It is an easily-applied model validation method (Ortúzar & Willumsen, 1999) and has been used in a few extant studies (Wong, Szeto, Wong, & Yang, 2014). FPR is the percentage of respondents who select the option with the highest modeled probability (or the percentage of choices correctly predicted by the model according to the maximum utility classification). FPR analysis compares the FPR with two indicators, namely chance recovery (CR) and expected recovery (ER), to confirm whether the model is both informative and reasonable.

CR is the percentage of the first preference choice provided by the equally probable model. The value and standard error of CR can be calculated as:

$$\mathrm{CR} = \frac{1}{N}\sum_i \frac{1}{M_i},$$

and

$$\mathrm{SE(CR)} = \frac{1}{N}\sqrt{\sum_i \frac{1}{M_i}\left(1 - \frac{1}{M_i}\right)},$$

where N is the number of validation observations and M_i is the number of satisfaction levels of the walking system experienced by respondent *i*.

ER is the expected proportion of FPR estimated from the model over the validation subsample. The value and standard error of ER can be calculated as:

$$\mathrm{ER} = \frac{1}{N}\sum_i P_i^{\max},$$

and

$$\mathrm{SE(ER)} = \frac{1}{N}\sqrt{\sum_i P_i^{\max}\left(1 - P_i^{\max}\right)},$$

where $P_i^{\max}$ is the estimated probability of the first preference option of respondent *i*.

Model validation has two hypotheses: (1) "there is no difference between the values of FPR and CR. If the test statistic exceeds the threshold value that is specified for the normal distribution at the chosen level of significance, we reject the hypothesis that the value of FPR is equal to that of CR and conclude that the model is informative"; and (2) "there is no difference between the values of FPR and ER. If the test statistic does not exceed the threshold value that is specified for the normal distribution at the chosen level of significance, we do not reject the hypothesis that the value of FPR is equal to that of ER and conclude that the model is reasonable." (Wong et al., 2014, p. 53).

Results

Model calibration

Our samples are randomly split into two subsamples, one for model estimation (517 observations) and the other for model validation (106 observations). The "hold-back" 106 observations were utilized to validate the OPM developed using 517 observations. Note that the proportion of calibration and validation subsample is in line with de D Ortúzar (1983) (= 600/724).

A pair-wise correlation analysis was performed to identify the association between independent variables. Independent variables are analyzed in pairs to determine their correlation. The results (shown in Tables A1 and A2) illustrate that multicollinearity does not occur in the data.

In addition to quantifying the contribution of each attribute, we aim to capture the travel system evolution over time. There are typically three alternatives for incorporating the time effect in the model (i.e. separate estimation, adding interaction terms, and adding a time dummy variable). Separate estimation and adding interaction terms do not work well in this study due to limited sample size and/or multi-collinearity. Hence, we add a time dummy variable into the model and estimate a single OPM using the pooled database with eleven explanatory variables. The model can be mathematically expressed as $y = f(A, T)$, where y is the dependent variable, A is a vector of service quality attributes, and T is a time dummy. As such, explanatory variables include ten service quality attributes and one dummy variable (which equals one for February 2018 data and zero for October 2017 data).

Eviews (v 9.0) was employed to calculate coefficients associated and the intercepts (or cut-off points). Table 4 reports the results of the OPM. Since no tourists reported the "highly dissatisfaction" choice on the walking system, the rating scale becomes four levels (ranging from 2 to 5). As such, three intercepts are separating the four choices.

The likelihood ratio test statistic (LR) is 161.48. As the number of explanatory variables is 11, this statistic has Chi-square distributed with the degree of freedom of 11 under the null hypothesis that none of the eleven variables has any effect. The p-value shown below LR is less than 0.001, indicating the strong overall significance of the developed OPM. The pseudo R^2, a widely-used goodness-of-fit measure, is 0.18, which suggests that the OPM fits the data well. Moreover, the three intercepts are significant at the 1% level. Lastly, the coefficient associated with the time dummy is positive, indicating that tourist satisfaction level increased from October 2017 to February 2018, controlling for other perceived environmental factors. This provides some evidence for the enhancement of the walking system during the interim.

Table 4 shows that eight out of ten service quality attributes are significant at the 10% level, except pavement and signage. The satisfaction levels of pavement and signage are not significant factors in explaining the difference in the overall satisfaction with the walking system. Observe that the two insignificant variables still maintain the expected sign (+). In addition, the level of crowding has the largest coefficient (0.34), followed by cleanliness (0.24). Furthermore, the estimated coefficient of each attribute is positive, which is in line with *a priori* expectations. It reveals that tourists who were more satisfied with each attribute had a better perception of the service itself.

Table 5 lists the marginal effects of all variables on four satisfaction levels. The marginal effects of each attribute on different satisfaction levels sum up to zero. The marginal effects of the level of crowding for the levels of high satisfaction, satisfaction, and neutral are 0.096, −0.057, and −0.039, respectively. This indicates that a one-unit increase in the satisfaction level of crowding corresponds to 9.6% more likely to be very satisfied with the overall performance, 5.7% less likely to be satisfied, and 3.9% less likely to be neutral. Note that among the concerned attributes, the level of crowding (0.096) exerts the highest marginal effect on high satisfaction level, followed by cleanliness (0.066). Furthermore, compared to the tourists surveyed in October 2017, the counterparts visiting the island in February 2018 were 12.1% more likely to be very satisfied with the overall performance of the walking system. This outcome might be a reflection of the enhancement of the walking system during the intervening months.

Model validation

Table 6 presents the results of model validation. The value of FPR equals 67.92%, indicating that 72 out of 106 respondents from the validation subsample selected the satisfaction level to which the developed OPM assigns the highest probability. This illustrates that our OPM demonstrates a good fit to the external (or validation) data.

FPR value lies more than 2.58 standard errors from the CR mean. Hence, the first null hypothesis (there is no difference between FPR and CR) is rejected, which confirms that our OPM is informative. Moreover, FPR value lies within 1.96 standard errors from the ER mean calculated from the developed OPM. Therefore, we have insufficient evidence to reject the second null hypothesis (there is no difference between FPR and ER). This outcome indicates that the developed OPM is reasonable. Hence, the model developed by 517 estimation observations also performs well on 106 validation observations. We can thus conclude that the OPM specification is both informative and reasonable and that can explain the variation in the data well. This also illustrates that the relation between the tourists' levels of satisfaction with individual service quality attributes and overall performance can be well described and modeled by the OPM specification.

Conclusions and implications

The walking system in traffic-free destinations is more important to tourists compared with that in destinations with other available travel modes. Hence, understanding how tourists perceive the quality of the walking system is the first and foremost step in providing an improved travel experience for tourists. We considered a customer view, investigated perceived environmental factors that affect the overall satisfaction with the walking system, and evaluated the relative importance of each service aspect. By using Kulangsu as an example and based on an OPM, we determined that in the view of tourists, the two most essential attributes of the walking system are the level of crowding and cleanliness. Notably, to our best knowledge, they have not been pointed out in related walkability literature. The FPR analysis of the model specification's out-of-sample predictive performance guarantees the robustness of our key findings. Accordingly, policy formulation should be concentrated on the two identified attributes of the walking system, namely the level of crowding and cleanliness. The findings enhance our understanding of tourist satisfaction with the walking environment. We believe that the proposed measures may assist transportation and tourism planners in formulating practical and appropriate policy measures for tourist destinations with similar travel conditions.

Table 4. OPM results for overall satisfaction with the walking system (N = 517).

Variable	Coefficient	Std. dev.	*z*-statistic
Convenience (e.g. few obstacles)	0.15[a]	0.09	1.69
Comfort	0.18[a]	0.09	1.87
The presence of shade (e.g. trees and arcade)	0.19[a]	0.08	2.24
Street lighting	0.21[a]	0.09	2.36
Pavement	0.01	0.09	0.07
Cleanliness	0.24[a]	0.10	2.33
The level of crowding	0.34[a]	0.08	4.06
Signage	0.04	0.08	0.54
Environmental art	0.22[a]	0.09	2.55
WIFI service	0.20[a]	0.07	2.83
Time dummy	0.43[a]	0.11	3.77
α_2 (intercept separating the levels of 2 and 3)	0.15[a]	0.09	1.69
α_3 (intercept separating the levels of 3 and 4)	0.18[a]	0.09	1.87
α_4 (intercept separating the levels of 4 and 5)	0.19[a]	0.08	2.24
Number of observations	517		
Log-likelihood of the null model		−446.35	
Maximized log-likelihood		−365.61	
LR		161.48	
p-value		<0.001	
Pseudo R^2		0.18	

a significant at the 10% level.

Table 5. OPM marginal effects of all significant quality attributes (N = 517).

	Marginal effects on			
Service quality attribute	dissatisfaction	neutral	satisfaction	high satisfaction
Convenience (e.g. few obstacles)	-6.39E-05	-0.017[a]	-0.025	0.042[a]
Comfort	-7.46E-05	-0.020[a]	-0.029[a]	0.049[a]
The presence of shade (e.g. trees and arcade)	-7.90E-05	-0.021[a]	-0.031[a]	0.052[a]
Street lighting	-8.77E-05	-0.023[a]	-0.034[a]	0.058[a]
Pavement	-2.46E-06	-0.001	-0.001	0.002
Cleanliness	-1.01E-04	-0.027[a]	-0.039[a]	0.066[a]
The level of crowding	-1.46E-04	-0.039[a]	0.057[a]	0.096[a]
Signage	-1.87E-05	-0.005	-0.007	0.012
Environmental art	-9.26E-05	-0.025[a]	-0.036[a]	0.061[a]
WIFI service	-8.58E-05	-0.023[a]	-0.033[a]	0.056[a]
Time dummy	-1.84E-04	-0.049[a]	-0.072[a]	0.121[a]

a significant at the 10% level.

Table 6. Model validation results (N = 106).

Measure	Result
Validation sample size	106
FPR (%)	67.92 (= 72/106)
CR (%)	25
2.58 standard errors (%)	10.85
Conclusion of CR hypothesis test[a]	Reject
ER (%)	66.45
1.96 standard errors (%)	8.82
Conclusion of ER hypothesis test[b]	Do not reject

[a]Null hypothesis tests at the 1% level; [b]Null hypothesis tests at the 5% level.

The level of crowding is substantially negatively perceived by Kulangsu tourists and holds a top priority for improvement. The level of crowding holds a top priority for improvement, which echoes the finding of Li et al. that this attribute is substantially negatively perceived by Kulangsu tourists. Decreasing the level of crowding appeals to the preferences and needs of the tourists. A host of surveyed tourists complained to us that at present, Kulangsu is considerably crowded, thereby decreasing their propensity and willingness to revisit (or repurchase) and recommend to friends/relatives (i.e. tourist loyalty). However, tourist volume would inevitably increase because of Kulangsu's inclusion on the World Heritage Sites list (see Figure 3). As such, decreasing the level of crowding appears to be extremely difficult. Thus, four suggestions are proposed. First, improving road connectivity to spread tourist demand evenly through the whole walking network can be explored. Second, unpopular scenic spots and areas can be vitalized to balance tourist distribution by quite a few measures. For example, destination marketing organizations can develop *ex nihilo* strategies (e.g. organizing music-themed and historical-anecdote-themed shows and creating artificial attractions) and construct adverts for unpopular areas. Third, the joint-ticket scheme, which has been long ignored but has the potential to effectively promote unpopular locations (Li et al., 2019), can be carefully designed and implemented. Lastly, given that tourist's length of stay has modestly reduced from the first to the second phase, destination marketing organizations can focus on encouraging tourists to spend more days on the island, which may optimize the tourism carrying capability and divert over-congestion (Masiero et al., 2018). We believe that such measures could effectively divert over-congestion (i.e. decrease the level of crowding or smoothing the peaks and troughs of tourist demand), promote a spatially balanced tourist distribution, and enhance the level of local services. Accordingly, a systematic enhancement method and accurate determination of priorities are necessary.

Cleanliness is another aspect worth improving. At present, a multitude of construction projects is implemented in Kulangsu, partially in response to its inclusion on the World Heritage Sites list. However, a few of these projects engender noticeable adverse spillover effects or side effects (e.g. noise, raising dust, air pollution, and unfavorable landscape), which consequentially ruin the quality of Kulangsu tourism. We observe that the level of satisfaction with cleanliness modestly decreased (see Table 3). Hence, reducing the nuisance effects of these works is suggested, such as establishing visual barriers through vegetation and creating acoustic barriers wherever appropriate. Moreover, the hygiene of the walking system is suggested to be periodically monitored and maintained, particularly in areas with numerous tourist clusters.

For other attributes, such as the presence of shade and comfort, the need for improvement is relatively low. The improvement of these quality attributes should only be pursued when the cost is low. Moreover, given that pavement and signage are not significant in the OPM, enhancement appears unnecessary, at least at the moment.

This study has several strengths. To the best of our knowledge, this research is the first to explain the satisfaction with the walking system in a traffic-free tourist destination using the "perceived" individual service quality attributes. In addition, we validated our developed model specification, which was determined to fit the external data well, thereby confirming its informativeness and reasonability. This finding ensures that the relative importance of quality attributes is relatively stable and robust. Furthermore, the two most valuable variables identified (e.g. the level of crowding and cleanliness) have received little investigation in extant literature, the majority of which is conducted in contexts of the USA, Europe, Australia. We conjecture that this is due in part to the difference in contexts of this study and existing literature. For example, the high level of crowding is usually not a problem in low-/medium-density settings (e.g. the USA and Australia) but may matter in high-density places (e.g. urban China). Additionally, the difference in contexts is further augmented by focusing on a very popular tourist destination. As such, this study contributes to limited research on walking perception in a high-density (and traffic-free) context.

Various areas must be explored further in upcoming studies. First, we assume that walking system satisfaction is homogeneous between tourists with different socioeconomic status and treat them as a homogenous group. However, the perception of service quality can change depending on the category of users being considered, as dell'Olio, Ibeas, and Cecín (2010) and Alonso, Barreda, dell'Olio, and Ibeas (2018) state. Tourists with different socioeconomic characteristics (e.g. age and gender) may have varying preferences of service quality. For example, the mobility of older populations is generally lower compared with that of their younger counterparts (Szeto, Yang, Wong, Li, & Wong, 2017; Yang, 2018) and older populations may value the level of crowding and personal safety more. With a sufficient sample size, we can adopt this analytic framework to study the perception of different categories of tourists. Put another way, we only investigated environmental factors that influence pedestrian satisfaction and disregarded other potential determinants that are related to tourists' socio-demographic attributes and travel profile. It should be noted that we also attempted to incorporate non-environmental factors into our OPM, though the majority of them are determined to be insignificant and that the modeling result cannot provide additional insights for policy intervention. Second, we used a method to identify the attributes that need the most improvement. However, this approach only uses subjective measures. Indeed, the use of objective measurement (e.g. path width and segment slope) is also important to obtain a comprehensive view of service quality. We can merge subjective and objective measures for an improved evaluation of service performance. Third, we only look at the perception of walking environmental quality of tourists rather than that of local residents or both. A comparison of the perceptions of different groups may be necessary. Lastly, OPM addresses the ordinal nature of the outcome variable. However, in this study, independent variables are also ordinal, but they are treated similarly with continuous variables by OPM. This issue is an inherent shortcoming of OPM. The nature of this study prevented us from attempting to devise new and sophisticated models to circumvent the limitation and make a significant methodological breakthrough.

Acknowledgments

This study was supported by grants from the National Natural Science Foundation (No.41671141), Fujian Natural Science Foundation (No. 2015J01226), College Research Foundation of Xiamen University (No. 20720170046), and Xiamen Science and Technology Bureau project (No. 3502Z20183005). The authors are grateful to the Guest Editor and the two reviewers for their constructive comments.

Funding

This work was supported by the Natural Science Foundation of Fujian Province [2015J01226]; Xiamen Science and Technology Bureau project [3502Z20183005]; College Funding of Xiamen University [20720170046]; National Natural Science Foundation of China [41671141].

ORCID

Linchuan Yang http://orcid.org/0000-0001-6070-9044
Guibo Sun http://orcid.org/0000-0001-8493-2953
Yuan Li http://orcid.org/0000-0002-6747-8549

References

Aitchison, J., & Silvey, S. D. (1957). The generalization of Probit analysis to the case of multiple responses. *Biometrika*, *44*(1–2), 131–140.

Alonso, B., Barreda, R., dell'Olio, L., & Ibeas, A. (2018). Modelling user perception of taxi service quality. *Transport Policy*, *63*, 157–164.

Anderson, E. W., Fornell, C., & Lehmann, D. R. (1994). Customer satisfaction, market share, and profitability: Findings from Sweden. *The Journal of Marketing*, *58*(3), 53–66.

Baker, D. A., & Crompton, J. L. (2000). Quality, satisfaction and behavioral intentions. *Annals of Tourism Research*, *27*(3), 785–804.

Ben-Akiva, M. E., & Lerman, S. R. (1985). *Discrete choice analysis: Theory and application to travel demand (Vol. 9)*. Cambridge, MA: MIT Press.

Bopp, M., Wilcox, S., Laken, M., Butler, K., Carter, R. E., McClorin, L., & Yancey, A. (2006). Factors associated with physical activity among African-American men and women. *American Journal of Preventive Medicine*, *30*(4), 340–346.

Bowie, D., & Chang, J. C. (2005). Tourist satisfaction: A view from a mixed international guided package tour. *Journal of Vacation Marketing*, *11*(4), 303–322.

Brownson, R. C., Hoehner, C. M., Day, K., Forsyth, A., & Sallis, J. F. (2009). Measuring the built environment for physical activity: State of the science. *American Journal of Preventive Medicine*, *36*(4), S99–S123.

Burton, N. W., Turrell, G., Oldenburg, B., & Sallis, J. F. (2005). The relative contributions of psychological, social, and environmental variables to explain participation in walking, moderate-, and vigorous-intensity leisure-time physical activity. *Journal of Physical Activity and Health*, *2*(2), 181–196.

Cao, X., Handy, S. L., & Mokhtarian, P. L. (2006). The influences of the built environment and residential self-selection on pedestrian behavior: Evidence from Austin, TX. *Transportation*, *33*(1), 1–20.

Cerin, E., Leslie, E., Du Toit, L., Owen, N., & Frank, L. D. (2007). Destinations that matter: Associations with walking for transport. *Health & Place*, *13*(3), 713–724.

Cervero, R., & Duncan, M. (2003). Walking, bicycling, and urban landscapes: Evidence from the San Francisco Bay Area. *American Journal of Public Health*, *93*(9), 1478–1483.

Choi, J., Kim, S., Min, D., Lee, D., & Kim, S. (2016). Human-centered designs, characteristics of urban streets, and pedestrian perceptions. *Journal of Advanced Transportation*, *50*(1), 120–137.

De Bourdeaudhuij, I., Sallis, J. F., & Saelens, B. E. (2003). Environmental correlates of physical activity in a sample of Belgian adults. *American Journal of Health Promotion*, *18*(1), 83–92.

De Bourdeaudhuij, I., Teixeira, P. J., Cardon, G., & Deforche, B. (2005). Environmental and psychosocial correlates of physical activity in Portuguese and Belgian adults. *Public Health Nutrition*, *8*(7), 886–895.

de D Ortúzar, J. (1983). Nested logit models for mixed-mode travel in urban corridors. *Transportation Research Part A: General*, *17A*(4), 283–299.

dell'Olio, L., Ibeas, A., & Cecín, P. (2010). Modelling user perception of bus transit quality. *Transport Policy*, *17*(6), 388–397.

Guo, Q, Xu, P, Pei, X, Wong, S. C, & Yao, D. (2017). The effect of road network patterns on pedestrian safety: a zone-based bayesian spatial modeling approach. *Accident Analysis & Prevention*, *99*, 114-124.

Hall, J., O'Mahony, B., & Gayler, J. (2017). Modelling the relationship between attribute satisfaction, overall satisfaction, and behavioural intentions in Australian ski resorts. *Journal of Travel & Tourism Marketing*, *34*(6), 764–778.

Hardy, A., Hyslop, S., Booth, K., Robards, B., Aryal, J., Gretzel, U., et al. (2017). Tracking tourists' travel with smartphone-based GPS technology: A methodological discussion. *Information Technology & Tourism*, *17*(3), 255–274.

Hoehner, C. M., Ramirez, L. K. B., Elliott, M. B., Handy, S. L., & Brownson, R. C. (2005). Perceived and objective environmental measures and physical activity among urban adults. *American Journal of Preventive Medicine*, *28*(2), 105–116.

Huh, J., Uysal, M., & McCleary, K. (2006). Cultural/heritage destinations: Tourist satisfaction and market segmentation. *Journal of Hospitality & Leisure Marketing*, *14*(3), 81–99.

Hui, T. K, Wan, D, & Ho, A. (2007). Tourists' satisfaction, recommendation and revisiting singapore. *Tourism Management*, *28*(4), 965-975. doi:10.1016/j.tourman.2006.08.008

Jensen, S. (2007). Pedestrian and bicyclist level of service on roadway segments. *Transportation Research Record: Journal of the Transportation Research Board, 2031*, 43–51.

Junio, M. M. V., Kim, J. H., & Lee, T. J. (2017). Competitiveness attributes of a medical tourism destination: The case of South Korea with importance-performance analysis. *Journal of Travel & Tourism Marketing, 34*(4), 444–460.

Kim, S., Choi, J., & Kim, S. (2013). Roadside walking environments and major factors affecting pedestrian level of service. *International Journal of Urban Sciences, 17*(3), 304–315.

Kim, S., Park, S., & Lee, J. S. (2014). Meso-or micro-scale? Environmental factors influencing pedestrian satisfaction. *Transportation Research Part D: Transport and Environment, 30*, 10–20.

Lee, S., Jeon, S., & Kim, D. (2011). The impact of tour quality and tourist satisfaction on tourist loyalty: the case of chinese tourists in korea. *Tourism Management, 32*(5), 1115-1124. doi:10.1016/j.tourman.2010.09.016

Leyden, K. M. (2003). Social capital and the built environment: The importance of walkable neighborhoods. *American Journal of Public Health, 93*(9), 1546–1551.

Li, Y., Xiao, L., Ye, Y., Xu, W., & Law, A. (2016). Understanding tourist space at a historic site through space syntax analysis: The case of Gulangyu, China. *Tourism Management, 52*, 30–43.

Li, Y., Ye, Y., Xiao, L., Xu, W., Law, A., & Wang, D. (2017). Classifying community space at a historic site through cognitive mapping and GPS tracking: The case of Gulangyu, China. *URBAN DESIGN International, 22*(2), 127–149.

Li, Y., Yang, L., Shen, H., & Wu, Z. (2019). Modeling intra-destination travel behavior of tourists through spatio-temporal analysis. *Journal of Destination Marketing & Management, 11*, 260–269.

Litman, T. (2003). Economic value of walkability. *Transportation Research Record: Journal of the Transportation Research Board, 1828*, 3–11.

Lu, Y., Gou, Z., Xiao, Y., Sarkar, C., & Zacharias, J. (2018). Do transit-oriented developments (tods) and established urban neighborhoods have similar walking levels in hong kong?. *International Journal Of Environmental Research and Public Health, 15*(3), 555.

Manson, J. E., Hu, F. B., Rich-Edwards, J. W., Colditz, G. A., Stampfer, M. J., Willett, W. C., … Hennekens, C. H. (1999). A prospective study of walking as compared with vigorous exercise in the prevention of coronary heart disease in women. *New England Journal of Medicine, 341*(9), 650–658.

Masiero, L., Qian, J., Fong, D., & Law, R. (2018). Gambling destinations and the effect of gambling results on tourist satisfaction and loyalty. *Journal of Travel & Tourism Marketing, 35*(5), 678–689.

Murphy, L., Moscardo, G., Benckendorff, P., & Pearce, P. (2011). Evaluating tourist satisfaction with the retail experience in a typical tourist shopping village. *Journal of Retailing and Consumer Services, 18*(4), 302–310.

Olawole, M. O., & Aloba, O. (2014). Mobility characteristics of the elderly and their associated level of satisfaction with transport services in Osogbo, Southwestern Nigeria. *Transport Policy, 35*, 105–116.

Oliver, R. L. (2014). *Satisfaction: A behavior perspective on the consumer*. London and New York: Routledge.

Orstad, S. L., McDonough, M. H., Stapleton, S., Altincekic, C., & Troped, P. J. (2017). A systematic review of agreement between perceived and objective neighborhood environment measures and associations with physical activity outcomes. *Environment and Behavior, 49*(8), 904–932.

Ortúzar, J. D. D., & Willumsen, L. G. (1999). *Modelling transport*. Chichester: John Wiley & Sons.

Oviedo-García, M. Á., Vega-Vázquez, M., Castellanos-Verdugo, M., & Reyes-Guizar, L. A. (2016). Tourist satisfaction and the souvenir shopping of domestic tourists: Extended weekends in Spain. *Current Issues in Tourism, 19*(8), 845–860.

Parahoo, S. K., Harvey, H. L., & Radi, G. Y. A. (2014). Satisfaction of tourists with public transport: An empirical investigation in Dubai. *Journal of Travel & Tourism Marketing, 31*(8), 1004–1017.

Reed, J. A., Wilson, D. K., Ainsworth, B. E., Bowles, H., & Mixon, G. (2006). Perceptions of neighborhood sidewalks on walking and physical activity patterns in a southeastern community in the US. *Journal of Physical Activity and Health, 3*(2), 243–253.

Saelens, B. E., & Handy, S. L. (2008). Built environment correlates of walking: A review. *Medicine and Science in Sports and Exercise, 40*(7 Suppl), S550–S564.

Shoval, N., & Isaacson, M. (2007). Tracking tourists in the digital age. *Annals of Tourism Research, 34*(1), 141–159.

Song, H., Van der Veen, R., Li, G., & Chen, J. L. (2012). The Hong Kong tourist satisfaction index. *Annals of Tourism Research, 39*(1), 459–479.

Speakman, C. (2005). Tourism and transport: Future prospects. *Tourism and Hospitality Planning and Development, 2*(2), 129–135.

Spence, J. C., Plotnikoff, R. C., Rovniak, L. S., Ginis, K. A. M., Rodgers, W., & Lear, S. A. (2006). Perceived neighbourhood correlates of walking among participants visiting the "Canada on the Move" website. *Canadian Journal of Public Health/Revue Canadienne De Sante'e Publique, 97*, S36–S40.

Spreng, R. A., MacKenzie, S. B., & Olshavsky, R. W. (1996). A reexamination of the determinants of consumer satisfaction. *The Journal of Marketing, 60*(3), 15–32.

Suminski, R. R., Poston, W. S. C., Petosa, R. L., Stevens, E., & Katzenmoyer, L. M. (2005). Features of the neighborhood environment and walking by US adults. *American Journal of Preventive Medicine, 28*(2), 149–155.

Sun, G., Han, X., Sun, S., & Oreskovic, N. (2018). Living in school catchment neighborhoods: Perceived built environments and active commuting behaviors of children in China. *Journal of Transport & Health, 8*, 251–261.

Szeto, W. Y., Yang, L., Wong, R. C. P., Li, Y. C., & Wong, S. C. (2017). Spatio-temporal travel characteristics of the elderly in an ageing society. *Travel Behaviour and Society, 9*, 10–20.

Tian, G., & Ewing, R. (2017). A walk trip generation model for Portland, OR. *Transportation Research Part D: Transport and Environment, 52*, 340–353.

Tosun, C., Temizkan, S. P., Timothy, D. J., & Fyall, A. (2007). Tourist shopping experiences and satisfaction. *International Journal of Tourism Research, 9*(2), 87–102.

Truong, T. H., & Foster, D. (2006). Using HOLSAT to evaluate tourist satisfaction at destinations: The case of Australian holidaymakers in Vietnam. *Tourism Management, 27*(5), 842–855.

Voorhees, C. C., Ashwood, J. S., Evenson, K. R., Sirard, J. R., Rung, A. L., Dowda, M., & McKenzie, T. L. (2010). Neighborhood design and perceptions: Relationship with active commuting. *Medicine and Science in Sports and Exercise, 42*(7), 1253.

Wang, B., & Loo, B. P. (2018). Travel time use and its impact on high-speed-railway passengers' travel satisfaction in the e-society. *International Journal of Sustainable Transportation*, 1–13. doi:10.1080/15568318.2018.1459968.

Wang, W., Li, P., Wang, W., & Namgung, M. (2011). Exploring determinants of pedestrians' satisfaction with sidewalk environments: Case study in Korea. *Journal of Urban Planning and Development*, *138*(2), 166–172.

Wong, R. C. P., Szeto, W. Y., Wong, S. C., & Yang, H. (2014). Modelling multi-period customer-searching behaviour of taxi drivers. *Transportmetrica B: Transport Dynamics*, *2*(1), 40–59.

Xu, W., & Yang, L. (2019). Evaluating the urban land use plan with transit accessibility, sustainable cities and society. *45*, 474–485.

Xu, X., Xie, S., Wong, S. C., Xu, P., Huang, H., & Pei, X. (2016). Severity of pedestrian injuries due to traffic crashes at signalized intersections in hong kong: a bayesian spatial logit model. *Journal Of Advanced Transportation*, *50*(8), 2015–2028.

Yang, L. (2018). Modeling the mobility choices of older people in a transit-oriented city: policy insights. *Policy Insights, Habitat International*, *76*, 10–18. doi:10.1016/j.habitatint.2018.05.007

Yang, L., Wang, B., Zhou, J., & Wang, X. (2018). Walking accessibility and property prices, transportation research part d. *Transport and Environment*, *62*, 551–562.

Ye, Y., Li, D., & Liu, X. (2018). How block density and typology affect urban vitality: an exploratory analysis in shenzhen, china. *Urban Geography*, *39*(4), 631–652.

Appendix

Table A1. Variable codes.

E1	Convenience (e.g. few obstacles)
E2	Comfort
E3	The presence of shade (e.g. trees and arcade)
E4	Street lighting
E5	Pavement
E6	Cleanliness
E7	The level of crowding
E8	Signage
E9	Environmental art
E10	WIFI service
E11	Time dummy

Table A2. Correlation matrix.

	E1	E2	E3	E4	E5	E6	E7	E8	E9	E10	E11
E1	1.00										
E2	0.38	1.00									
E3	0.28	0.36	1.00								
E4	0.15	0.20	0.22	1.00							
E5	0.19	0.22	0.13	0.28	1.00						
E6	0.27	0.32	0.17	0.28	0.35	1.00					
E7	0.18	0.27	0.24	0.28	0.33	0.29	1.00				
E8	0.29	0.23	0.21	0.26	0.24	0.22	0.33	1.00			
E9	0.18	0.23	0.19	0.26	0.23	0.32	0.21	0.27	1.00		
E10	0.14	0.15	0.17	0.25	0.16	0.16	0.30	0.24	0.22	1.00	
E11	-0.09	-0.09	-0.01	-0.01	-0.09	-0.05	0.00	0.01	0.04	0.21	1.00

"I deserve a break!": how temporal landmarks and the perception of deservingness influence consumers' travel motivation and intention

Qing Li and Larry Yu

ABSTRACT

This study investigates the impact of temporal landmarks on consumers' travel motivation and intention through three experimental studies. Results demonstrate that being reminded of public and personal temporal landmarks can boost consumers' travel motivation, with the effect driven by consumers' perception of deservingness. The effect of temporal landmarks on travel motivation was shown to be weakened when consumers were confronted with goal conflicts attributed internally to consumers themselves, but not those attributed externally to others. We provide evidence that perceptions of deservingness influence travel intention, and these perceptions are influenced by both temporal landmarks and goal conflicts that are internally attributed.

Introduction

In today's fast-paced work environment, many workers forgo some of their annual vacation days due to work conflicts or competitive pressures in the workplace, which raises concerns for employee well-being and work productivity. The annual Vacation Deprivation Study by Expedia (2017) revealed a work–life imbalance in many countries, such as the United States (US), Japan, South Korea, and Mexico. These vacation-deprived countries, as described in the Expedia (2017) study, have fewer paid vacation days compared with the vacation-moderate and vacation-rich countries. Moreover, workers in the vacation-deprived countries also have one thing in common: they do not fully spend the vacation days they earn. The Japanese had 10 unused vacation days in 2016; the South Koreans, seven; and the Americans, three. Because of this phenomenon of vacation deprivation, the tourism industry and business organizations have advocated the importance of fully using paid vacation days for travel and leisure to improve personal well-being (U.S. Travel Association, 2017).

Travel for enjoyment is determined by many factors, such as a desire to temporarily leave one's residence, leisure time, and disposable income. Travel motivation is then triggered by many factors traditionally categorized as push and pull factors (Dann, 1977; Gnoth, 1997). The push comes from one's self, one's family, or one's workplace, such as a significant event in one's life, a desire to go to a specific place, a desire to recover from work stress, or simply a desire to use paid vacation time. The pull forces come from the tourism destinations amplified by multiple channels of distribution, such as destination marketing in both traditional and digital outlets, social media, and electronic word-of-mouth communication, to influence and motivate potential consumers to make travel decisions.

A plethora of studies has examined travel motivation (Dann, 1981; Figler, Weinstein, & Sollers, 1992; Fodness, 1994; Mansfeld, 1992; Pearce & Caltabiano, 1983; Pearce & Lee, 2005). Hsu and Huang (2008) provided a comprehensive review of current research on travel motivation, including different conceptual approaches and methodological issues. Studies have considered tourists of different nationalities, demographics, and types, such as travel motivations of German tourists (Jamrozy & Uysal, 1994), travel motivations of Taiwanese seniors (Jang & Wu, 2006), comparative analysis of travel motivations of British, American, and Canadian tourists of different genders to visit Barbados (Jönsson & Devonish, 2008), the cultural effect on travel motivation of Chinese outbound tourists (Li & Cai, 2012), travel motivations of Japanese outbound tourists (Cha, McCleary, & Uysal, 1995), and the factors motivating Canadian ecotourists (Eagles, 1992).

Current research on tourism motivation has examined the physiological, behavioral, cognitive, and social aspects of factors influencing individuals to make travel decisions. However, a research gap remains as to when and why individuals are more likely to decide on leisure travel, particularly in terms of the transmitting mechanism and boundary conditions influencing motivation when temporal landmarks are present and consumers perceive deservingness and goal conflict at work. In this paper, we explored individuals' perceptions and experiences that heightened or reduced motivation for leisure travel and proposed that individuals develop enhanced motivation for leisure travel by recognizing temporal landmarks (both salient public and significant personal landmarks) and perceiving deservingness; this motivation can be decreased by goal conflict, particularly when consumers attribute the conflict internally to themselves rather than others.

Conceptual background and hypotheses

Temporal landmarks

Temporal landmarks refer to significant dates that have special meaning for individuals, such as public holidays, birthdays, graduations, and weddings. Temporal landmarks are typically described in two categories: public or social temporal landmarks that everyone shares, such as national holidays, festivals, or the beginning of a new year, and personal temporal landmarks that are unique to individuals, such as birthdays, weddings, anniversaries, and achievements such as receiving a major award or promotion, or reaching a career milestone (Dai, Milkman, & Riis, 2014). In studying personal histories using autobiographical recall of events, Shum (1998, p. 425) defined three basic types of personal temporal landmarks: public events as flashbulb memories, personal events associated with one's first experiences, and extended events determined by reference points in calendars. These temporal landmarks have been applied to examine autobiographical recall of calendar events and effect (Kurbat, Shevell, & Rips, 1998; Shum, 1998). For example, the study by Kurbat et al. (1998) found that students recalled more events at the beginning and end of semesters than at other times. Shum (1998) reported that temporal landmarks were identified as effective cues for autobiographical recall and dating accuracy by individuals.

In recent years, more studies have examined the significance of temporal landmarks for self-appraisal between current and future selves (Peetz & Wilson, 2013, 2014) and the influence of temporal landmarks on motivating task/goal initiation (Dai, Milkman, & Riis, 2015; Tu & Soman, 2014) and aspirational behavior (Dai et al., 2014). Furthermore, temporal landmarks have been shown to be selectively used by individuals to "regulate the experienced inclusion or exclusion of a temporally extended self" (Peetz & Wilson, 2014, p. 53). It has been shown that when dealing with a negative future self, people use holidays as temporal landmarks to demarcate the current and future self, yet when considering a positive future self, people seldom choose temporal landmarks as an intervening factor to separate the present and future self (Peetz & Wilson, 2014, p. 53).

Several other empirical studies have shown temporal landmarks as motivational factors for goal or task initiation. For example, in five studies, Tu and Soman (2014) examined whether consumer task initiation was affected by how consumers determined task deadlines using temporal landmarks, for example, the last day of the year, the last day of the month, or an important dinner. Their findings showed that consumers tended to initiate a task when the deadline was defined in "a like-the-present category" rather than in "an unlike-the-present category" (Tu & Soman, 2014, p. 819). Other studies have also found a causal relationship between temporal landmarks marking the beginning of a new time period and people's increased intention to initiate goal pursuit (Dai et al., 2015, p. 1927), and this causal effect of temporal landmarks on motivating aspirational behavior was described as the "fresh start effect" (Dai et al., 2014, p. 2563).

Although many temporal landmarks are associated with holidays and vacations, such as the end of the school year ushering in the summer travel season or the national vacation policy on institutionalizing numerous temporal landmarks, such as traditional festivals, as public holidays in China (Zhang, Singh, & Yu, 2016), the tourism research literature has not addressed how temporal landmarks motivate people's desire to travel and celebrate their life events or milestone achievements and have a fresh start after holiday or vacation travel.

Perception of deservingness

Perception of deservingness has been extensively studied in political psychology and social psychology to examine either positive or negative outcomes for self or others in regard to actions (Feather, 1999), such as how public policy on aid to the poor is influenced by perception of deservingness (Applebaum, 2001), the perception of deservingness among citizens in European welfare states (Van Oorschot, 2006), the perception of deservingness in equity distribution and promotions in organizations (Feather, 2008), and resource allocation for health care patients (Skinner, Feather, & Freeman, 2007). In business, perception of deservingness has been rigorously promoted in marketing to remind consumers to purchase products and service. Studies have found that perception of deservingness influences consumers' propensity to purchase (Taylor, Webb, & Sheeran, 2014).

This concept has been recently applied to the fields of consumer behavior and marketing research, although businesses have utilized the concept in advertising for decades – for example, "you deserve a break today" by McDonald's since 1971. Scholars have been studying deservingness and indulgence consumption as consumers are motivated to get what they deserve. One study revealed that advertisements focusing on deservedness were more effective for people with a strong belief in a just world mediated by the degree to which the advertisements matched participants' world view and justified buying the product, but were less effective for high-knowledge participants (Hafer, Mantonakis, Fitzgerald, & Bogaert, 2016). Research has also found that sending consumers reminders of valued relationships with family and friends can actually reduce their perception of deservingness and restrict indulgent consumption if those relationships are no longer strong (Cavanaugh, 2014, p. 218). In their study of price fairness, Ashworth and McShane (2012) found that a violation of deservingness was an effective mediator between consumer comparisons of product prices and seller unfairness. This burgeoning field of research on consumer cognition provides an insight into purchasing decisions for indulgent consumptions such as leisure travel.

Goal conflict and attributions of goal conflict

Numerous studies have revealed that goal conflict, particularly conflict at a multiple or higher level of abstraction, negatively affects individuals' psychological well-being (Kehr, 2003; Gray, Ozer, & Rosenthal, 2017; Kelly, Mansell, & Wood, 2015). Goal conflict occurs when a person has several goals that cannot all be pursued at the same time (Boudreaux & Ozer, 2013); conflict occurs when pursuing one goal starts to interfere with other goals (Fishbach & Ferguson, 2007). Goal conflict is a largely subjective experience; individuals have different perceptions of the degree of conflict they experience, and the conflict may seem greater on some occasions than others (Etkin, Evangelidis, & Aaker, 2015, p. 395; Laran & Janiszewski, 2009). A question of substantive interest in the goal conflict literature is the degree to which this conflict is associated with behavioral inhibition.

Two types of goal conflict are identified in the literature: limited resources, defined as resource goal conflict due to time, money, and energy; and incompatible strategies for goal attainment, known as inherent goal conflict, when pursuing one goal inhibits the pursuit of another goal (Segerstrom & Nes, 2006). Gorges and Grund (2017, p. 9) further stated that "inherent goal conflict occurs when individuals take on goals set by other individuals (or goals that are at least shaped by expectations of significant others) – for example, at work, within families – this type of goal conflict should be very relevant for individuals' lives as well." Most of the studies on goal conflict have focused on resource goal conflict (Gorges & Grund, 2017). For example, is time an inhibiting factor for consumers facing goal conflicts? One study found that when consumers faced goal conflicts due to time and other demands, they felt more stress and then felt that they had less time to pursue personal goals (Etkin et al., 2015, p. 394). As Etkin et al. (2015, p. 395) noted, "Perceiving greater conflict between goals has an impact on experienced emotions," and heightened goal conflict increases consumers' stress and anxiety levels. However, few studies have been devoted to inherent goal conflict due to the complex nature of the types of goal conflict. Gorges and Grund (2017, p. 9) called for more investigations into inherent goal conflict and comparison of "antecedents, experiential instances, and consequences" with those of resource goal conflict as "an important yet insufficiently engaged task."

Goal conflicts are therefore attributed to resource constraints and inherent conflicts by work-related materialistic values and personal-related social values (Gorges & Grund, 2017). Goal conflicts can occur due to individual internal goal motivation and external organizational goal priority and complexity. Such conflicts can also result from a need to balance goals associated with materialistic values and social values, such as meeting work assignments requested by management and giving up or reducing time for leisure travel. To investigate the perceived causes of goal conflict, we applied attribution theory, in which causation of the conflict is perceived as resulting from one's own behavior or other persons' behavior (Kelley & Michela, 1980). Attribution therefore refers to "the cognitive processes through which an individual infers the cause of an actor's behavior" (Calder & Burnkrant, 1977, p. 29). Internal causes for a conflict include personal or dispositional factors, and external causes include situational factors (Thibaut & Riecken, 1955). Many studies have examined the attributions of perceived consumer behaviors, such as interpersonal influence (Calder & Burnkrant, 1977) and consumers' reaction to negative word-of-mouth communication (Laczniak, DeCarlo, & Ramaswami, 2001). However, few studies have investigated attributions of goal conflict in consumer perceptions and behaviors. One exception is the study of Carlson, Kacmar, and Stepina (1995), who examined attributions of work–family conflict facing data processing professionals. This study revealed that internal causes had a stronger impact on conflict than external causes; the reason was that individuals tended to blame themselves instead of attributing the conflict to external causes out of their control (Carlson et al., 1995, p. 24). The purpose of this study was to fill the theoretical void by probing attributions of goal conflict and the impact of the perceived cause on perception of deservingness and travel motivation and intention.

Travel motivation and intention

Travel motivation is the key driving force behind travel behavior and is typically triggered by the push and pull factors that influence consumers to form attitudes toward travel products and services (Dann, 1981). Motivation is considered an antecedent of behavior intention, as shown in Lewin's (1951, pp. 95–96) definition of a complete intentional action: "Its first phase is a motivational process, whether a brief or protected vigorous struggle of motives; the second phase is an [mental] act of choice, decision, or intention, terminating this struggling; the third phase is the consummatory intentional action itself." Behavioral intention can thus lead to purchase decisions for travel and can serve as a valid proxy for actual behavior, where measures would be inappropriate or difficult to obtain (Lu, Hung, Wang, Schuett, & Hu, 2016). It is clear that both travel motivation and intention influence the final purchasing behavior of consumers. We therefore examined both constructs to reveal the impact of temporal landmarks on consumers' travel behavior. Methodologically, both constructs were used to enhance the robustness and the external validity of the study.

Figure 1 provides an overview of the studies and the proposed conceptual model for this research. This moderated mediation model aimed to test the linkage between temporal landmarks and travel motivation and intention through the intervening mechanism of perception of deservingness and the boundary condition of attributions of goal conflict between temporal landmarks and perceived deservingness. Guided by the conceptual model, we developed and tested research hypotheses by conducting three experimental studies.

Hypotheses

The impact of temporal landmarks on travel motivation

As previous studies have proposed, temporal landmarks can be a marker for separating one's present self and future self (Peetz & Wilson, 2014), it can serve as a fresh start (Dai et al., 2014), and can help categorize time (Tu & Soman, 2014). Highlighting the scarcity of that time may heighten individuals' perception of deservingness by focusing their thoughts on pursuing something indulgent that they have wanted to do but have never had a reason to do (i.e. leisure travel). Furthermore, temporal landmarks can be defined in a like-the-present category or an unlike-the-present category (Tu & Soman, 2014). That is, the temporal landmark is the dividing point between past and future. Individuals may feel that they deserve to indulge themselves because of the hard work that they completed in the past or because of the hard work that will commence in the future,

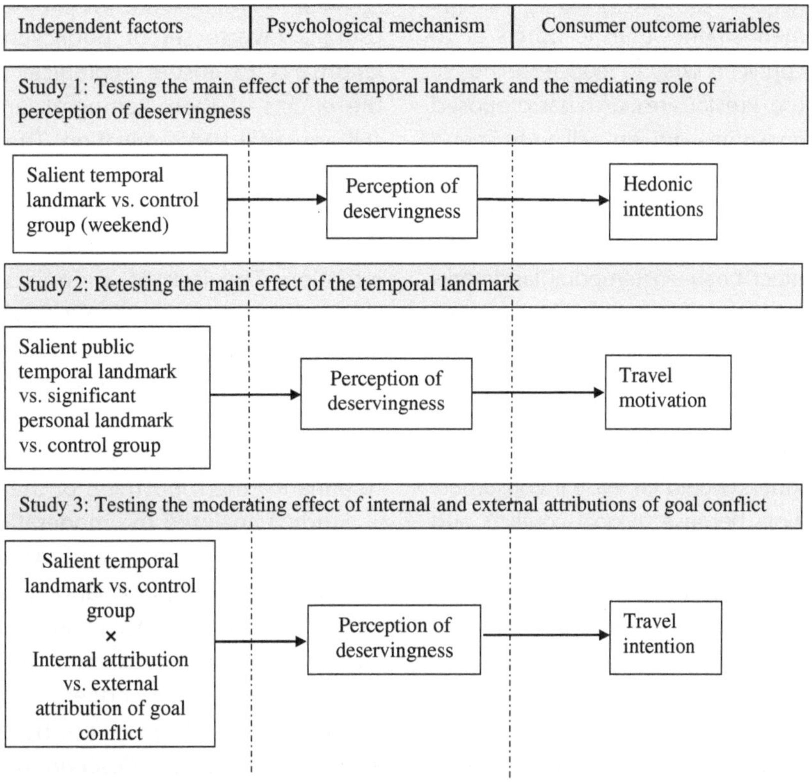

Figure 1. Overview of the study and conceptual model.

since perception of deservingness is shown as consumers believe they earn the right to indulge through hard work (Kivetz & Simonson, 2002; Mick & Demoss, 1990; Pereira-Heath, Tynan, & Ennew, 2011). This theorizing is captured by the following hypotheses :

H1: Consumers are more likely to decide to travel when reminded of a salient temporal landmark (significant public landmark or personal landmark) rather than an ordinary or mundane day.

H2: The effect proposed in H1 occurs because consumers perceive themselves to be more deserving of pleasure (leisure travel) on a salient temporal landmark (significant public landmark or personal landmark) than on an ordinary or mundane day.

Moderating effect of goal conflict

Many workers face a work–pleasure conflict due to the balance of competitive pressures and family pressures in today's fast-paced work environment. Boudreaux and Ozer (2013) demonstrated that individuals with a greater conflict were less successful in attaining goals that were not necessary. It is widely believed that pleasure is less necessary than work. When consumers face goal conflicts due to time and other demands, they abandon their pursuit of personal pleasure (Etkin et al., 2015, p. 394). Particularly when they attribute the goal conflict to their own ability to manage time for work requirements, rather than to others or to situational factors, the conflict is likely to reduce feelings of deservingness of pleasure. Previous research has proposed that time-constrained consumers are less willing to spend time on travel-related activities (Etkin et al., 2015, p. 394). Accordingly, we predicted that internal attributions of goal conflicts that make people feel pressed for time would moderate the causal effect between temporal landmarks and perception of deservingness on travel motivation. Salient temporal landmarks remind consumers of and inspire hedonic consumption of leisure travel, and such motivation is further facilitated by perception of deservingness of such indulgent consumption. However, the level of perception of deservingness would increase if consumers felt fewer time constraints because of goal conflicts and would be decreased if they felt more time constraints due to goal conflicts, especially when the conflicts were internally attributed or self-inflicted. This leads to the following hypothesis:

H3: The effect of temporal landmarks on consumer travel motivation is diminished when consumers encounter goal conflicts that are internally attributed, rather than externally attributed.

Overview of the research

As Figure 1 illustrates, we conducted three experiments to test the relationships proposed in the model: temporal landmarks' direct influence on travel motivation, its indirect effect on travel motivation through the intervening mechanism of perception of deservingness, and the moderating effect of goal conflcit attributions on the relationships of temporal landmarks, perception of deservingness and travel motivation. Study 1 contrasted the direct effects of a specific temporal landmark (university anniversary) with an ordinary or mundane day on travel motivation and intention, and then the mediating effect of perception of deservingness between the temporal landmark and travel motivation.

In order to enhance the robustness of the study, Study 2 examined different types of temporal landmarks, salient public temporal landmarks versus significant personal landmarks. If there is a difference in perception of specificity between the two, that influence needs to be excluded, thereby further verifying the mediating role of perception of deservingness. Study 2 therefore replicated the first study but broadened the context of temporal landmarks to contrast salient public temporal landmarks and significant personal temporal landmarks. Previous studies on temporal landmarks examined both salient public temporal landmarks and significant personal landmarks. For example, Peetz and Wilson (2013) conducted five studies by using a diverse set of public and personal temporal landmarks to ensure ecological validity in examining the effects of these temporal landmarks on temporal self-appraisal and motivation. Therefore, Study 2 in our experiment compared the difference between the two types of temporal landmarks to determine if they had significantly different effects on travel motivation and intention. The lack of a significant difference would imply that a temporal landmark encompasses a wide range of temporal events and, therefore, would ensure the external validity of the study by determining robust effects across the two types of temporal landmarks. In general, Study 2 expanded the externalities of Study 1 to enhance the robustness of the research.

Study 3 analyzed the moderating effect of goal conflict attributed internally and externally on the relationship between temporal landmarks and perception of deservingness. The test of attribution of goal conflict was used to further examine the mediating effect of perception of deservingness. We hypothesized that facing conflict of goals attributed to oneself would weaken perception of deservingness and thus the effect of the temporal landmark on travel motivation and intention. However, if the goal of conflict was attributed externally, such as an inherent goal conflict set by others, perception of deservingness would still exist, and the effect of the temporal landmark on travel motivation and intention would not be affected. By examining the moderating effect of goal conflict attribution, the perception of deservingness could be further verified as an effective mediator, enhancing the robustness of the research. Taken together, these three studies provide empirical evidence to test the hypotheses that temporal landmarks, both public and personal, influence travel motivation directly as well as indirectly through perception of deservingness, and that the effect of temporal landmarks on perception of deservingness is moderated by internal attribution of goal conflict.

Study 1

The objective of Study 1 was to test our prediction that a salient temporal landmark (versus an ordinary or mundane day) can positively change consumers' hedonic intentions and that this relationship is positively affected by the intervening mechanism of perception of deservingness. The university anniversary was used as the salient temporal landmark for this experiment since it was jubilantly celebrated and students had a day off on this special day; a regular Saturday was used as the control. We then examined the mediating effect of perception of deservingness on the relationship of the salient temporal landmarks and hedonic intention and eliminated another explanation. Due to the way in which the temporal landmark condition was created in this study, it is possible that our results were driven by the perceived importance of the day. Therefore, to examine whether consumers' perception of importance of the day might have contributed to our results, we included one item that measured this construct.

Sample and procedures

Participants comprised 51 undergraduate junior and senior students (28 women; M_{age} = 21.3 years) who were recruited from a management course at a large university in Southern China and received course credit in return for their participation. They were told that they would complete a study about college life and were assigned to one of two conditions: salient temporal landmark or the control group condition. This study was conducted two days before the one-hundred-and-tenth anniversary of the university and three days before the weekend. For the salient temporal landmark, we first reminded the participants of the university anniversary and then asked them how important they thought the university anniversary was to evaluate the importance of this anniversary. For the control group, participants were reminded of the weekend and were asked the same question (i.e. how important they thought the weekend was) to evaluate the importance of the weekend. They were asked to report their hedonic intentions on the anniversary (versus on the weekend), and the experiment was conducted in Chinese. At the end of the experiment, participants were debriefed and thanked for their participation.

Measures

Perception of deservingness was measured by using a 4-item scale validated in Cavanaugh's (2014) study (α = 0.924). We followed the process of back-translation recommended by Brislin (1986) to first convert the English items into Chinese and then translate the Chinese version back into English. The back-translated English items were compared with the original English ones by a marketing professor for semantic consistency. Minor discrepancies were noted and we modified the Chinese version accordingly. The four statements were "I feel I deserve to reward myself," "I feel I deserve to treat myself to nice things," "I feel I deserve to indulge myself a little," and "I feel I deserve to buy something special for myself." Participants rated all responses on a 7-point scale (1 = "strongly disagree," and 7 = "strongly agree"). For Study 1, to measure hedonic intention with the same 7-point scale, we adopted three statements used by Lu et al. (2016) (α = 0.945): "I might travel on that special day," "I will plan to travel on that particularly day," and "I am willing (intend) to travel on that particular day." The consumers' perception of the importance of the day was measured by first describing the university anniversary to the participants and then asking them: "How important do you think this day is to you?" This question also applied a 7-point scale, with 1 as least important and 7 as most important.

Results and discussion

An analysis of variance (ANOVA) revealed a main effect of temporal landmarks ($F[1, 48] = 4.731$, $p = 0.035 < 0.05$ [$\eta^2 = 0.09$]), showing that participants' hedonic intentions for indulgence on the salient temporal landmark day, the university anniversary ($M_{School\ Day} = 5.40$), were higher than those on the weekend ($M_{Weekend} = 3.98$). Then we used Hayes's (2013) PROCESS macro for SPSS (model 4) to conduct a mediation analysis to explore if perception of deservingness and perception of importance of the day mediated the effect of the temporal landmark on hedonic intentions. A bootstrapping test using 5000 resamples at a

95% confidence interval for perception of deservingness showed a lower-limit confidence interval (LLCI) of −1.8298 and an upper-limit confidence interval (ULCI) of −0.1962 that did not include zero, confirming that perception of deservingness mediated the relationship between temporal landmarks and hedonic intentions. However, no mediation effect of perception of importance of the day was found: university anniversary day (M = 4.45; SD = 1.96) and weekend (M = 3.81; SD = 1.14), with *t*-test result of $t(49) = 1.489$, $p = 0.144$. This result was further verified by the 95% bootstrap confidence intervals for the mediation test that included zero (95% LLCI = −0.929, ULCI = 0.044; 5,000 resamples). Thus, a salient temporal landmark (versus an ordinary day) had a more positive effect on consumers' hedonic intention by increasing consumers' perceived deservingness, supporting H1.

Study 1 revealed that salient temporal landmarks influenced consumers' intention for hedonic consumption more than a regular day, such as a regular Saturday. A salient temporal landmark motivates consumers to celebrate special or significant events and also open their wallet for a treat to celebrate the special event and have a good time. Furthermore, the causal effect of salient temporal landmarks on hedonic intention was heightened by consumers' perception of deservingness, as they believed that the special event was the right time for self-indulgence. Such indulgent consumption intention could be found for other types of salient temporal landmarks, including end-of-school parties, public holidays, recurring and non-recurring festivals, religious holidays, traditional festivals of different cultures, and mega sporting events. These results therefore exhibit how salient temporal landmarks influence hedonic consumption intention and provide empirical support for the transmitting mechanism of perception of deservingness. The next study, Study 2, replicated this study but included significant temporal landmarks in an individual consumer's life, such as a birthday, significant achievement, or anniversary.

Study 2

In Study 2, we replicated the temporal landmark effects by contrasting a salient public temporal landmark with a significant personal landmark. We further elucidated the mechanism underlying these observed patterns and eliminated some other explanations. Due to the way in which the temporal landmark condition was created in Study 2, it is possible that our results were driven by the perceived specificity of the day. Therefore, to examine whether consumers' perception of specificity of the day might have contributed to our results, we included one item that measured this construct.

Sample and procedures

We recruited a sample of 167 participants (101 women; $M_{age} = 19.8$ years) enrolled in a marketing course from a large university in Southern China. The participants were juniors and seniors in the business school and they completed this study in exchange for course credit. They were randomly assigned to one of three experimental conditions, according to a 1 × 3 (temporal landmark: salient public temporal landmark versus significant personal landmark versus control group) between-subjects design. All participants were told that the study was about planning leisure activities and completed this study in one large classroom. First, they were asked to report a specific day that is important to the public (salient public temporal landmark, e.g. public holiday or traditional festival) or significant to themselves (significant personal landmark, e.g. birthday) or a weekend (control group, e.g. Saturday) and evaluate the specificity of that day by answering the question: "How specific do you think that day is?" Participants were then asked to report their perceptions of deservingness and then complete a 25-item measurement of travel motivation. Once the participants had completed all required tasks individually, they were debriefed and thanked for their participation. The experiment was conducted in Chinese.

Measures

Replicating Study 1, the Chinese version of Cavanaugh's (2014) 4-item scale of perception of deservingness ($\alpha = 0.924$) was used in this study, and each question was measured on a 7-point scale (1 = "strongly disagree," and 7 = "strongly agree"). The 25-item measurement of travel motivation was adopted from Lu et al. (2016, $\alpha = 0.904$). The same process of back-translation recommended by Brislin (1986) was followed as in Study 1, for semantic consistency and cultural relevancy in the translated Chinese questionnaire. Items included "I like to broaden my views," "I want to be excited," "I want to experience the joy of relaxation," and "I want to escape the routine of my life." The study by Lu et al. (2016) found that specific events may significantly influence the travel motivations of tourists, resulting in different travel intentions. The same study suggested that these events strongly impact the process of purchasing consumer goods and the type of travel and tourism products and services consumed.

Results and discussion

ANOVA revealed a mean effect of temporal landmarks ($F[2, 164] = 10.76$, $p < 0.001 < 0.01$ [$\eta^2 = 0.116$]). Participants' travel motivations for a salient public temporal landmark were higher than for a weekend, t(123) = 4.216, $p < 0.001$, as were their travel motivations for a significant personal landmark, t(98) = 3.52, $p = 0.001$. However, participants' travel motivations did not differ for the public versus personal temporal landmark, t(107) = 0.173, $p = 0.863$.

We next conducted a mediation analysis using the PROCESS macro (Hayes, 2013) to test our hypothesis that the effect of temporal landmarks on travel motivation was driven by perception of deservingness, but not perception of specificity. The results of this analysis showed evidence of mediation through perception of deservingness (95% LLCI = − 0.267, ULCI = −0.0825; 5000 resamples), but not perception of specificity (95% LLCI = − 0.1384, ULCI = 0.0038; 5000 resamples).

Study 2 replicated findings from Study 1 by testing the difference between salient public temporal landmarks and significant personal landmarks, or the influence of specificity, to re-examine the mechanism of temporal landmark effect on travel motivation and intention. Findings revealed that salient public temporal landmarks had a main effect on travel motivation. Additionally, we found that personal significant landmarks also had an influence on consumers' motivation/intention for leisure activities such as travel. There was a significant difference in the perception of specificity between temporal landmarks and ordinary or mundane days. Therefore, while there was no evidence that a regular weekend increased consumers' motivation for travel consumption and there was a difference in perception of specificity between salient public temporal landmark and significant personal landmark, there was no significant effect of the two types of temporal landmarks on travel motivation and intention, implying that perception of specificity was not an intervening mechanism. Clearly, temporal landmarks, both salient public temporal landmarks and significant personal temporal landmarks, can influence leisure travel motivations. Furthermore, the intervening mechanism of perception of deservingness provided a significant mediating effect between temporal landmarks, both public and personal, and leisure travel motivations. Would the effect of temporal landmarks on leisure travel motivation mediated by perception of deservingness be decreased if consumers were confronted with goal conflicts that resulted in time constraints? This question was examined in Study 3.

Study 3

Study 3 sought to test the moderating effect of internal and external attributions of goal conflict on travel motivation and intention through the mediated effect between temporal landmark and perception of deservingness. We examined whether goal conflict attributed internally and externally attenuates the effect of the temporal landmark on leisure travel motivation.

Sample and procedures

Our sample consisted of 87 undergraduate sophomores and junior students enrolled in a marketing class (69% females; M_{age} = 19.9 years) at a business school of a large university in Southern China. The participants were mostly business majors and completed this study in exchange for course credit. They were randomly assigned to one of four conditions in a 2 (temporal landmark: salient temporal landmark versus weekend) × 2 (goal conflict: internal attribution versus external attribution) between-subjects design. Participants first were asked to report a specific day that is important to the public (salient public temporal landmark, e.g. public holiday, traditional festival) or a weekend (control group, e.g. Saturday), which was similar to Study 2. They were then asked to read a brief description of a goal conflict attribution scenario, in which the conflict was caused internally, due to the participants themselves, or caused externally, due to others. The scenario of internal attribution of goal conflict described that, on the day before the salient temporal landmark or the day before Saturday, the professor reminded the class that a paper would be due three days after the temporal landmark or Saturday. The student realized that the paper was assigned a month ago and he or she completely forgot about it. The scenario of external attribution of goal conflict described that, on the day before the salient temporal landmark or the day before Saturday, the professor assigned the class new homework that would be due three days after the temporal landmark or Saturday. The participants were asked to report their travel intention in light of goal conflicts attributed internally or externally. All participants were debriefed and thanked for their participation after completing the experiment.

Measures

In addition to the 4-item perception of deservingness measure by Cavanaugh (2014), we used three statements in this study to measure travel intention by Lu et al. (2016) (α = 0.945) on a 7-point scale ranging from 1, "strongly disagree," to 7, "strongly agree." These statements were "I might travel on that special day;" "I will plan to travel on that particular day;" and "I am willing (intend) to travel on that particular day." These three questions were also used in Study 1 for measuring travel intention.

Results and discussion

A two-way ANOVA analysis revealed a significant interaction between temporal landmarks and goal conflict attribution ($F[1, 83] = 4.71$, $p = 0.033 < 0.05$ [$\eta^2 = 0.054$]), as expected. When the goal conflict was perceived as externally attributed, participants were significantly more willing to travel on the public holiday, the salient temporal landmark ($M = 3.88$), versus the weekend ($M = 2.69$; $t[40] = 2.59$, $p = 0.013 < 0.05$). When the goal conflict was perceived as internally attributed, participants' travel intention was not significantly different for the salient temporal landmarks ($M = 2.47$) or the weekend ($M = 2.59$; $t[43] = -0.304$, $p = 0.763$), as illustrated in Figure 2.

Study 3 provided evidence for our prediction that consumers' perception of deservingness would decrease if they faced goal conflicts that they had caused. The conflict between meeting work deadlines and leisure activity affected consumer travel intention due to time constraints. When consumers juggle multiple goals in their work and life with limited time, they have to attend to the most important goal that will affect their work performance or family life, and leisure pursuits could yield to these priorities and be postponed. However, it is interesting to note that the moderating effect of goal conflict was insignificant if the conflict was attributed to others. In this case, the school project was assigned so late and the time allotted for completing it was so short because of the oversight of the instructor. Therefore, this finding further emphasizes the conditional effect of goal conflict attribution affecting the influence of temporal landmarks on travel intention through the transmitting mechanism of perception of deservingness.

General discussion

Drawing on a consumer cognitive perspective, this study demonstrates the influential impact that temporal landmarks, both salient public temporal landmarks and significant personal landmarks, can have on perception of deservingness and hedonic consumption intention, and how goal conflict attributions, internal or external, can moderate the perception of deservingness. First, through a study of undergraduate management students, we found that temporal landmarks can motivate consumers to travel and enjoy leisure time by escalating consumers' perception of deservingness. Consumers are more likely to be hedonic with regard to a salient temporal landmark event, such as a public holiday, than on an ordinary weekend because they feel they deserve it more at that time rather than because they feel the salient temporal landmark is more important than a weekend (Study 1). Study 2 demonstrated that both a salient public temporal landmark and a significant personal landmark could inspire consumers' travel motivation, but their effects were not significantly different. This study re-examined the mediating role of consumers' perception of deservingness and excluded the explanation of perception of specificity. Study 3 explored the boundary condition for the relationship between salient temporal landmarks and perception of deservingness and found that when consumers are confronted with goal conflicts that they perceive as internal attribution, they feel less deserving, and the effect of salient temporal landmarks on their travel intention is eliminated or decreased. Conversely, the effect of salient temporal landmarks on consumers' travel intention is stronger if goal conflict is perceived as external attribution. The cumulative findings across the three studies provide converging evidence suggesting that salient temporal landmarks influence consumers' perception of deservingness, which increases hedonic intention, which, in turn, is constrained by goal conflicts consumers attribute to themselves.

We conducted the experiments by deploying several designs and using different types of temporal landmarks (salient public temporal landmarks and significant personal landmarks), and we measured multiple outcomes related to leisure travel (hedonic intention, travel motivation, and travel intention). These approaches support the robustness and generalizability of our findings on how perceptions of deservingness influence travel intention and emphasize the important influence that temporal landmarks have on consumer travel motivation/intention.

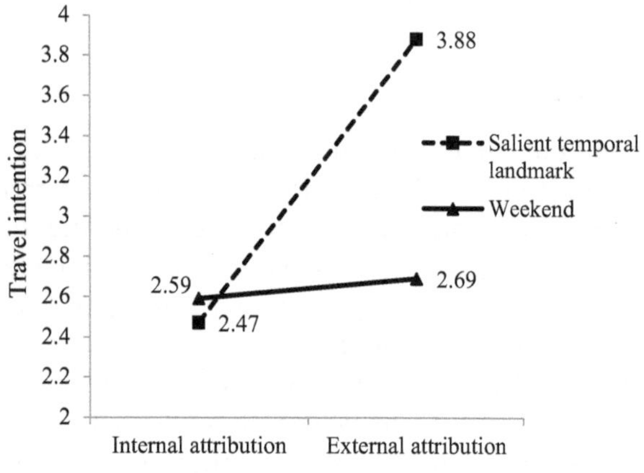

Figure 2. Travel intention in response to goal conflict attribution.

Theoretical implications

Our results have several theoretical implications for research on consumer perception and travel motivation. First, we theoretically contribute to the burgeoning research on temporal landmarks (Dai et al., 2014, 2015; Peetz & Wilson, 2013; Tu & Soman, 2014) by demonstrating that temporal landmarks can lead consumers to be hedonic and pursue leisure activities such as travel. Previous studies have focused on the impact of temporal landmarks to serve as a fresh point for self-appraisal and aspirational behavior (Dai et al., 2014; Peetz & Wilson, 2013), stimulate goal initiation (Dai et al., 2015), and organize future points in time for task implementation (Tu & Soman, 2014). Previous research proposed that temporal landmarks can motivate aspirational behaviors, such as goal pursuit (Dai et al., 2014, 2015). However, this research found that temporal landmarks can motivate hedonic behaviors, such as leisure travel, as well. Our research builds on the temporal landmarks theory by suggesting that salient public temporal landmarks and significant personal landmarks can spur consumer travel motivation and hedonic consumption. The causal effect of temporal landmarks in influencing consumer behaviors has been clearly established in our research.

Second, we demonstrate that consumers' perception of deservingness underlies the impact of temporal landmarks on consumers' travel motivation. Deservingness has been promoted in advertising by various industries for decades. Consumer studies have explored the perception of deservingness in several strands of literature, including consumer self-gifts (Mick & Demoss, 1990; Mick & Faure, 1998; Tynan, Pereira-Heath, Ennew, Wang, & Sun, 2010), justification for indulgence (Taylor et al., 2014), and perception of deservingness (Cavanaugh, 2014; Feather & McKee, 2009). We were able to provide support for the mediating role of perception of deservingness in the pathway from temporal landmarks to travel motivation – one of the first attempts of its kind in the tourism consumption literature. Perception of deservingness has an impactful influence on consumer hedonic intention to travel.

Third, our research spotlights the circumstances under which a reduced perception of deservingness decreases travel intention. We answered the call by Gorges and Grund (2017, p. 9) to investigate the moderating effect of goal conflict. In doing so, we applied attribution theory to test the moderating effect of internal and external causes of goal conflict on the mediating effect of perception of deservingness influencing travel intention. Few studies have determined the specific attributions of goal conflict influencing consumers' travel intention. Our research shows that goal conflict attribution has different effects on perception of deservingness depending on whether it is perceived to have an internal cause or an external cause. If consumers attribute the conflict to themselves, they will prioritize work-related goals and delay leisure-related activities. However, if they attribute the conflict to others as an inherent goal conflict, they will downplay the importance of work-related goals and continue to focus on leisure-related activities.

Finally, our study sheds new light on the research on travel motivation, particularly from the push aspect of motivation. Past studies on travel motivation have examined various influencing determinants such as time, disposable income, demographics, and pull factors driven by destination marketing, local travel service advertising, word-of-mouth referrals, and consumer-generated comments in social media. Our research suggests that consumer perceptions of hedonic consumption open a new avenue for the study of tourism motivation and behaviors.

Our results also have practical implications for tourism management, consumers, and public policy. The temporal landmark has significant practical implications for travel agency managers, hotel and resort managers, and hedonic product sellers. Hotels and resorts may best appeal to consumers' desire for leisure travel by advertising salient temporal landmarks and significant personal landmarks. Across the three studies in our research, creating temporal landmarks to motivate consumers to travel or go on vacation (Peetz & Wilson, 2013, p. 264) has been identified as an impactful factor. Though tourism destinations and businesses have been leveraging salient public temporal landmarks, significant personal temporal landmarks, and even ordinary weekends to promote business, they need to consider enhancing consumers' perception of deservingness to indulge in a well-deserved holiday, a celebration, or a special event by leveraging multiple channels of communication, advertisements, direct customer relationship management, and social media outlets. Heightened perception of deservingness for hedonic consumption increases consumers' intention to act. Advertising messages and customer relationship management should therefore include a reminder of deservingness for hedonic consumption.

Consumers face both resource goal conflict and inherent goal conflict between work and family, or work and life. Taking on multiple goals with limited time will increase stress and anxiety, which will affect goal attainment. Consequently, internal goal conflicts caused by consumers themselves can reduce their intention for hedonic consumption at temporal landmark events. Many of those in vacation-deprived countries, as identified by Expedia (2017), have even missed or sacrificed their deserved vacation time as part of the social phenomenon described as "work martyrs" or "under-vacationed America" in the US (U. S. Travel Association, 2017). This phenomenon also raises questions on how public policy can facilitate a mechanism for balancing goals between work and leisure in vacation-deprived countries where the "work martyrs" phenomenon is particularly prevalent. Business organizations should also formulate a plan to encourage employees to take their deserved time off for different temporal landmark events since these temporal landmarks can influence self-renewal and inspire individuals to have a new start (Dai et al., 2014).

Limitations and future research

These theoretical contributions and practical implications should be viewed in light of the study limitations. First, we selected a university anniversary as the salient temporal landmark and we selected a significant personal temporal landmark for influencing hedonic consumption intention. The selection of temporal landmarks in paradigms is determined by many factors (Peetz & Wilson, 2014). Using a variety of temporal landmarks may have had a different effect on consumers' hedonic intentions. Additionally, our studies focused only on individuals in their junior and senior years in college. The use of college students in social science research is recognized as leading to more homogeneous responses (Peterson, 2001). Consumer perceptions and behaviors may change as individuals age, and these changes influence social perceptions. As such, future research should consider replicating the research on non-student populations for generalization.

Future research should also explore other boundary conditions that influence the level of perception of deservingness and hedonic intentions, such as the moderating effects of personal and situational characteristics. For example, in this study we utilized goal conflict attribution to examine the effect of perception of deservingness on travel intention on an individual basis. Future examination of goal conflict attribution can be expanded to other inherent causes, such as work delayed by other team members at the team level. Additional studies could better simulate realistic consumer perceptions and behaviors to increase ecological validation.

Disclosure statement

No potential conflict of interest was reported by the authors.

Funding

This work was supported by the Pengcheng Scholar Program sponsored by Shenzhen Municipal Education Commission.

References

Applebaum, L. D. (2001). The influence of perceived deservingness on policy decisions regarding aid to the poor. *Political Psychology*, *22*(3), 419–442.

Ashworth, L., & McShane, L. (2012). Why do we care what others pay? The effect of other consumers' prices on influences of seller (dis)respect and perceptions of deservingness violation. *Journal of Retailing*, *88*(1), 145–155.

Boudreaux, M. J., & Ozer, D. J. (2013). Goal conflict, goal striving, and psychological well-being. *Motivation and Emotion*, *37*(3), 433–443.

Brislin, R. W. (1986). The wording and translation of research instruments. In W. J. Lonner & J. W. Berry (Eds.), *Field methods in cross-cultural research* (pp. 137–164). Beverly Hills, CA: Sage.

Calder, B. J., & Burnkrant, R. E. (1977). Interpersonal influence on consumer behavior: An attribution theory approach. *Journal of Consumer Research*, *4*(1), 29–38.

Carlson, D. S., Kacmar, K. M., & Stepina, L. P. (1995). An examination of two aspects of work-family conflict: Time and identity. *Women in Management Review*, *10*(2), 17–25.

Cavanaugh, L. A. (2014). Because I (don't) deserve it: How relationship reminders and deservingness influence consumer indulgence. *Journal of Marketing Research*, *51*, 218–232.

Cha, S., McCleary, K. W., & Uysal, M. (1995). Travel motivations of Japanese travelers: A factor-cluster segmentation approach. *Journal of Travel Research*, *34*(1), 33–39.

Dai, H. C., Milkman, K. L., & Riis, J. (2014). The fresh start: Temporal landmarks motivate aspiration behavior. *Management Science*, *60*(10), 2563–2582.

Dai, H. C., Milkman, K. L., & Riis, J. (2015). Put your imperfections behind you: Temporal landmarks spur goal initiation when they signal new beginnings. *Psychological Science*, *26*(12), 1927–1936.

Dann, G. M. S. (1977). Anomie, ego-enhancement and tourism. *Annals of Tourism Research*, *4*(4), 184–194.

Dann, G. M. S. (1981). Tourist motivation: An appraisal. *Annals of Tourism Research*, *8*(2), 187–219.

Eagles, P. F. J. (1992). The travel motivations of Canadian ecotourists. *Journal of Travel Research*, *31*(2), 3–7.

Etkin, J., Evangelidis, I., & Aaker, J. (2015). Pressed for time? Goal conflict shapes how time is perceived, spent, and valued. *Journal of Marketing Research*, *52*(3), 394–406.

Expedia. (2017). *Expedia 2017 vacation deprivation report*. Retrieved from https://viewfinder.expedia.com/news/millennials-vacation-deprived/.

Feather, N. T. (1999). Judgments of deservingness: Studies in the psychology of justice and achievement. *Personality and Social Psychology Review*, *3*(2), 86–107.

Feather, N. T. (2008). Perceived legitimacy of a promotion decision in relation to deservingness, entitlement, and resentment in the context of affirmative action and performance. *Journal of Applied Social Psychology*, *38*(5), 1230–1254.

Feather, N. T., & McKee, I. R. (2009). Differentiating emotions in relation to deserved or undeserved outcomes: A retrospective study of real-life events. *Cognition and Emotion*, *23*(5), 955–977.

Figler, M. H., Weinstein, A. R., & Sollers, J. J. (1992). Pleasure travel (tourist) motivation: A factor analysis approach. *Bulletin of the Psychonomic Society*, *30*(2), 113–116.

Fishbach, A., & Ferguson, M. J. (2007). The goal construct in social psychology. In A. W. Kruglanski & E. T. Higgins (Eds.), *Social psychology: Handbook of basic principles* (pp. 490–515). New York, NY: Guilford.

Fodness, D. (1994). Measuring tourist motivation. *Annals of Tourism Research*, *21*(3), 555–581.

Gnoth, J. (1997). Tourism motivation and expectation formation. *Annals of Tourism Research*, *24*(2), 283–304.

Gorges, J., & Grund, A. (2017). Aiming at a moving target: Theoretical and methodological considerations in the study of intraindividual goal conflict between personal goals. *Frontiers in Psychology*, *8*, 1–14.

Gray, J. S., Ozer, D. J., & Rosenthal, R. (2017). Goal conflict and psychological well-being: A meta-analysis. *Journal of Research in Personality*, *66*, 27–37.

Hafer, C. L., Mantonakis, A., Fitzgerald, R., & Bogaert, A. F. (2016). The effectiveness of deservingness-based advertising messages: The role of product knowledge and belief in a just world. *Canadian Journal of Administrative Science, 35* (1), 34–46.

Hayes, A. F. (2013). *Introduction to mediation, moderation, and conditional process analysis—A regression-based approach*. New York, NY: Guilford.

Hsu, C. H. C., & Huang, S. (2008). Travel motivation: A critical view of the concept's development. In A. G. Woodside & D. Martin (Eds.), *Tourism management: Analysis, behavior, and strategy* (pp. 14–27). Cambridge, MA: CABI.

Jamrozy, U., & Uysal, M. (1994). Travel motivation variations of overseas German visitors. In E. Kaynak & M. Uysal (Eds.), *Global tourist behavior* (pp. 135–160). Binghamton, NY: Haworth.

Jang, S. C., & Wu, C. M. E. (2006). Seniors' travel motivation and the influential factors: An examination of Taiwanese seniors. *Tourism Management, 27*(2), 306–316.

Jönsson, C., & Devonish, D. (2008). Does nationality, gender, and age affect travel motivation? A case of visitors to the Caribbean island of Barbados. *Journal of Travel & Tourism Marketing, 25*(3–4), 398–408.

Kehr, H. M. (2003). Goal conflicts, attainment of new goals, and well-being among managers. *Journal of Occupational Health Psychology, 8*(3), 195–208.

Kelley, H. H., & Michela, J. L. (1980). Attribution theory and research. *Annual Review of Psychology, 31*, 457–501.

Kelly, R. E., Mansell, W., & Wood, A. M. (2015). Goal conflict and well-being: A review and hierarchical model of goal conflict, ambivalence, self-discrepancy and self-concordance. *Personality and Individual Difference, 85*, 212–229.

Kivetz, R., & Simonson, I. (2002). Earning the right to indulge: Effort as a determinant of customer preferences toward frequency program rewards. *Journal of Marketing Research, 39*(2), 155–170.

Kurbat, M. A., Shevell, S. K., & Rips, L. J. (1998). A year's memories: The calendar effect in autobiographical recall. *Memory & Cognition, 26*(3), 532–552.

Laczniak, R. N., DeCarlo, T. E., & Ramaswami, S. N. (2001). Consumers' responses to negative word-of-mouth communication: An attribution theory perspective. *Journal of Consumer Psychology, 11*(1), 57–73.

Laran, J., & Janiszewski, C. (2009). Behavioral consistency and inconsistency in the resolution of goal conflict. *Journal of Consumer Research, 35*, 967–984.

Lewin, K. (1951). *Field theory in social science*. New York, NY: Harper and Row.

Li, M., & Cai, L. (2012). The effects of personal value on travel motivation and behavioral intention. *Journal of Travel Research, 51*(4), 473–487.

Lu, J., Hung, K., Wang, L., Schuett, M. A., & Hu, L. (2016). Do perceptions of time affect outbound-travel motivations and intention? An investigation among Chinese seniors. *Tourism Management, 53*, 1–12.

Mansfeld, Y. (1992). From motivation to actual travel. *Annals of Tourism Research, 19*(3), 399–419.

Mick, D. G., & Demoss, M. (1990). Self-gifts: Phenomenological insights from four contexts. *Journal of Consumer Research, 17*(1), 322–332.

Mick, D. G., & Faure, C. (1998). Consumer self-gifts in achievement contexts: The role of outcomes, attributions, emotions, and deservingness. *International Journal of Research in Marketing, 15*, 293–307.

Pearce, P. L., & Caltabiano, M. L. (1983). Inferring travel motivation from travelers' experiences. *Journal of Travel Research, 22*(2), 16–20.

Pearce, P. L., & Lee, U. (2005). Developing the travel career approach to tourist motivation. *Journal of Travel Research, 43* (3), 226–237.

Peetz, J., & Wilson, A. E. (2013). The post-birthday world: Consequences of temporal landmarks for temporal self-appraisal and motivation. *Journal of Personality and Social Psychology, 104*(2), 249–266.

Peetz, J., & Wilson, A. E. (2014). Marking time: Selective use of temporal landmarks as barriers between current and future selves. *Personality and Social Psychology Bulletin, 40*(1), 44–56.

Pereira-Heath, M. T., Tynan, C., & Ennew, C. T. (2011). Self-gift giving: Understanding consumers and exploring brand messages. *Journal of Marketing Communications, 17*(2), 127–144.

Peterson, R. A. (2001). On the use of college students in social science research: Insights from a second-order meta-analysis. *Journal of Consumer Research, 28*, 450–461.

Segerstrom, S. C., & Nes, L. S. (2006). When goals conflict but people prosper: The case of dispositional optimism. *Journal of Research in Personality, 40*(5), 675–693.

Shum, M. S. (1998). The role of temporal landmarks in autobiographical memory processes. *Psychological Bulletin, 124* (3), 423–442.

Skinner, N., Feather, N. T., & Freeman, T. (2007). Stigma and discrimination in health-care provision to drug users: The role of values, affect, and deservingness judgments. *Journal of Applied Social Psychology, 37*(1), 163–186.

Taylor, C., Webb, T. L., & Sheeran, P. (2014). 'I deserve a treat!': Justifications for indulgence undermine the translation of intentions into action. *British Journal of Social Psychology, 53*(3), 501–520.

Thibaut, J. W., & Riecken, H. W. (1955). Some determinants and consequences of the perception of social causality. *Journal of Personality, 22*(2), 113–133.

Tu, Y. P., & Soman, D. (2014). The categorization of time and its impact on task initiation. *Journal of Consumer Research, 41*, 810–822.

Tynan, C., Pereira-Heath, M. T., Ennew, C., Wang, F., & Sun, L. (2010). Self-gift giving in China and the UK: Collectivist versus individualist orientations. *Journal of Marketing Management, 26*, 1112–1128.

U.S. Travel Association. 2017. *Project: Time off*. Washington, DC: Author. Retrieved from. https://www.ustravel.org/programs/project-time

Van Oorschot, W. (2006). Making the difference in social Europe: Deservingness perceptions among citizens of European welfare states. *Journal of European Social Policy, 16*(1), 23–42.

Zhang, C., Singh, A. J., & Yu, L. (2016). Does it matter? Examining the impact of China's vacation policies on domestic tourism demand. *Journal of Hospitality & Tourism Research, 40*(5), 527–556.

Towards the recovery mechanisms of leisure travel experiences: does the length of vacation matter?

Chun-Chu Chen, Suiwen (Sharon) Zou and Jie Gao

ABSTRACT

This research examines the recovery mechanisms of leisure travel experiences with a focus on the moderating role of vacation duration. Results from a sample of 509 US respondents shows that autonomy is a requisite in the link between tourism experiences and life satisfaction, while its effect on life satisfaction is mediated by passive and active tourism experiences. Another key finding is that shorter leisure trips tend to facilitate the passive route to life satisfaction, while longer trips can better facilitate the active route. These findings provide important insights regarding the connection between tourism and well-being.

Introduction

Vacation is a leave of absence from the daily routine, which takes an individual away to a place far from usual surroundings. As a result, vacation offers a chance to rest and recover from everyday life demands and stress, and in turn, can increase life satisfaction. The connection between tourism experiences and well-being is well-documented in the literature (Uysal, Sirgy, Woo, & Kim, 2016). For instance, both Nawijn, Marchand, Veenhoven, and Vingerhoets (2010) and Gilbert and Abdullah (2004) found that the holiday-taking group experienced a higher level of life satisfaction than the non-holiday-taking group. In addition, de Bloom et al. (2009) conducted a meta-analysis and reported that vacation does have a positive impact on tourists' life satisfaction.

Although the positive effects of vacation on well-being are well documented, research on *how* vacation experiences affect well-being is still a significant void and lacks theoretical foundation (de Bloom et al., 2009; Lehto, 2013). It remains under-researched regarding how different vacation features (e.g. active versus passive vacation experiences, long versus short vacations) exert influence on life satisfaction (de Bloom et al., 2009). With the understanding of this underlying process, tourism business can better design and create a vacation experience that will result in improved subjective well-being (Fritz & Sonnentag, 2006). Thus, it is deemed important to uncover how specific vacation experiences affect life satisfaction.

Growing evidence suggested that taking a vacation may serve as a recovery experience that leads to an increase in life satisfaction (Fritz & Sonnentag, 2006; Chen, Petrick & Shahvali, 2016). Recovery can generally be defined as "a process during which individual functional systems that have been called upon during a stressful experience return to their pre-stressor levels" (Sonnentag & Fritz, 2007, p. 205). Taking time off for vacations has been suggested to increase life satisfaction through recovery experiences in two ways (Chen, Huang, & Petrick, 2016).

First, while the lack of autonomy is a major contributor to stress (Tennant, 2001), individuals may regain their autonomy after hours, at weekends or during vacations as they will be more likely to control how to use their time (Sonnentag & Fritz, 2007), and thus recover from the autonomy regaining process. Second, evidence suggests that passive experiences (such as detachment from work) and/or active experiences (such as participating in activities that provide learning or challenging opportunities) in leisure time can help buffer stress and reduce the negative impacts of the stress (Newman, Tay, & Diener, 2014), which contributes to individuals' recovery and life satisfaction.

Built on existing studies pertaining to work stress relief in leisure (Caldwell, 2005; Iwasaki, Zuzanek, & Mannell, 2001) and organizational behavior studies (Kinnunen et al., 2016; Sonnentag & Fritz, 2007), the overall purpose of this study is to uncover the psychologically recovery mechanisms of leisure travel experiences. In other words, this study looks at how leisure travel can help individuals recover from work. Specifically, there are two-fold objectives: (1) to develop and test a model of recovery process that includes perceived life satisfaction (after a leisure trip) as the outcome variable predicted by passive and active vacation experiences; and (2) to investigate how the process of recovery during a leisure trip is moderated by the duration of leisure travel.

Literature review

Work & vacation

Work stress is a growing concern for the employers because high work stress can lead to lower job performance and satisfaction, higher turnover rates, and absenteeism (Briner & Reynolds, 1999). Therefore, various corporations and organizations have utilized interventions to help their employees recover from a stressful working environment as an attempt to improve organizational performance (Combs, Liu, Hall, & Ketchen, 2006). These organizational interventions can be either directly associated with job design and human resources policies or related to resources for stress coping (Briner & Reynolds, 1999). Given that the benefits of leisure travel have been well-documented in the tourism literature (Chen & Petrick, 2013), providing paid vacation days for employees should be an effective intervention strategy. Unfortunately, workers in certain regions and countries are provided with far less paid vacation days.

According to Expedia Vacation Deprivation Study (Expedia, 2017), workers in most countries in the regions of North America (including: United States, Canada, and Mexico) and East Asia (including: South Korea, Hong Kong, Malaysia, Thailand, Singapore, and Japan) had approximately 15 paid vacation days in 2016, while their counterparts in most European countries on average earned more than 25 paid vacation days (see Table 1). More interestingly, workers in European countries managed to use nearly all their paid vacation days; however, workers in most North American and East Asian countries wasted a lot of paid vacation days (Expedia, 2017). Therefore, paid vacations seem to be a human right in Europe but a luxury in North America and East Asia.

Moreover, the statistics from the Organization for Economic Co-operation and Development, (2017) also show that workers in North America and East Asia had much longer working hours in 2016 as compared to the annual working hours in Europe. As shown in Table 1, the annual working hours for workers in most European

Table 1. Work-and-life balance indicators for selected countries.

Region/Countries	Vacation days[a] earned	Vacation days[b] taken	Annual working[c] hours	Life satisfaction[d]
Europe				
France	30	30	1,472	6.4
Spain	30	30	1,695	6.4
Finland	30	30	1,653	7.4
Germany	30	28	1,363	7.0
Italy	30	25	1,730	5.8
Austria	25	25	1,601	7.1
Denmark	25	25	1,410	7.5
Norway	25	25	1,424	7.6
Sweden	25	25	1,621	7.3
United Kingdom	25	25	1,676	6.5
Switzerland	25	25	1,590	7.6
Belgium	24	23	1,551	6.9
Netherlands	25	22	1,430	7.3
Ireland	21	21	1,879	6.8
Pacific				
New Zealand	20	19	1,752	7.4
Australia	20	15	1,669	7.3
North America				
Canada	15	14	1,703	7.4
USA	15	12	1,783	6.9
Mexico	15	12	2,255	6.2
East Asia				
Singapore	18	14	N/A	N/A
Hong Kong	14	14	N/A	N/A
Malaysia	16	12	N/A	N/A
Thailand	15	12	N/A	N/A
Japan	20	10	1,713	5.9
South Korea	15	8	2,069	5.8

[a,b]The numbers were drawn from Expedia Vacation Deprivation Study http://blog.expedia.co.uk/vacation-deprivation-2016/

[c,d]The numbers were drawn from OECD http://stats.oecd.org/index.aspx?DataSetCode = BLI

countries were in the range of 1,300 to 1,600 in 2016, while in North America and East Asia, workers on average worked for more than 1,800 hours in the same year. As workers in North America and East Asian have longer working hours and fewer vacation days, they miss out on opportunities for recovery at work. It is thus important to examine how leisure travel can help workers recover from the demands of daily work, particularly for workers in North America and East Asia. Work stress sets up the study background for this paper and motivates this study to examine the recovery role and mechanisms of leisure travel in relieving work stress.

Tourism & well-being

This study addressed the topic of tourism and well-being, which has gained increasing attention in the tourism literature (Lehto, 2013; Uysal et al., 2016). Generally, previous studies have shown that taking a leisure trip can lead to greater happiness and life satisfaction (Dolnicar, Yanamandram, & Cliff, 2012; Nawijn, Mitas, Lin, & Kerstetter, 2013; Sirgy, Kruger, Lee, & Yu, 2011). It is worth noting that health and wellness tourism has been identified as a niche market consists of people who travel mainly for the purpose of promoting health and well-being (Kim, Chiang, & Tang, 2017; Kim, Lee, & Ko, 2016), while research on tourism and well-being has demonstrated that everyone can benefit from taking a leisure trip (Chen et al., 2016).

This linkage between tourism experiences and the well-being of tourists is founded on the bottom-up theory of well-being (Diener, 1984), which suggests that an individual's well-being is an accumulation of pleasant and unpleasant life experiences. Research has shown that taking a leisure trip often time can help people gain a better appreciation for different aspects of life, such as work, marriage, leisure, or family, resulting in a more positive evaluation of life in general (Sirgy et al., 2011). However, people may not have a pleasant experience for every leisure trip. For example, if people are forced to stay connected to work when they are on vacation, their vacation experiences may be ruined by work demands (Chen, Huang, Gao, & Petrick, 2017).

The potential benefits of tourism experiences for specific populations have also been examined. For example, Wei and Milman (2002) have found that staying active during a leisure trip help elderly tourists feel happier and healthier. Research has also shown that tourism is an important means of increasing self-esteem and confidence for elderly tourists (Morgan, Pritchard, & Sedgley, 2015). Moreover, McCabe and his colleagues (McCabe & Johnson, 2013; McCabe, Joldersma, & Li, 2010) have demonstrated that providing free vacations for low-income families can help them feel higher levels of happiness and optimism, contributing to benefits in their family lives and personal relationships.

Even though previous studies have provided evidence of travel benefits for different populations, the connection between tourism and well-being was found unsupported in a few studies in organizational behavior (Fritz & Sonnentag, 2006; Tarumi, Hagihara, & Morimoto, 1998). Moreover, travel benefits are found to fade out quickly after the trip (Chen & Petrick, 2013). Several studies assessed respondents' perceived happiness and life satisfaction several times before and after leisure travel (Etzion, 2003; Kühnel & Sonnentag, 2011; Nawijn et al., 2010; Westman & Eden, 1997). Empirical findings suggest that travel benefits last for up to two months (Chen & Petrick, 2013), while most studies indicate that these benefits persist for less than two weeks (de Bloom et al., 2011; Etzion, 2003; Nawijn et al., 2010; Westman & Eden, 1997).

For full-time workers, it is important not only to understand the benefits of vacations, but also to know *how* to plan their vacations (e.g. vacation duration, type of activities) to maximize the vacation benefits and result in effective recovery. Even though the positive association between tourism and well-being are well documented, it remains under-researched regarding *how* vacation experiences stimulate the recovery process for full-time workers, which may lead to increased life satisfaction (Chen et al., 2016). Thus, research on this topic can provide helpful information to educate the public with the best practices of vacation planning backed by scientific evidence.

The recovery mechanisms of leisure travel

Many theoretical frameworks and concepts have been used to explain and explore the relationship between vacation and well-being, including attention-restoration theory (Lehto, 2013), the bottom-up theory (Neal, Uysal, & Sirgy, 2007), goal theory (Sirgy, 2010) and conservation of resources theory (Westman & Eden, 1997). This study is drawn on the recovery experience literature. Recovery is commonly defined as a process during which individuals work on their functional systems under stress to return to their pre-stressor levels (Sonnentag & Fritz, 2007). It is considered as a voluntarily maintained lifestyle characterized by sobriety, personal health, and citizenship (Belleau et al., 2007). Individuals gain new resources through recovery, which may lead to increased well-being and higher job performance (Fritz & Sonnentag, 2006). Thus, this research intended to uncover the underlying mechanisms of recovery and examine how full-time workers can recover through leisure travel experiences. Based on existing evidence in organizational behavior (Sonnentag & Fritz, 2007, 2015) and leisure studies (Iwasaki et al., 2001; Caldwell, 2005; Trenberth & Dewe, 2002), four psychological components of the recovery mechanisms have been identified in this study.

The first mechanism identified in the literature is autonomy, which can be defined as the level of control that a person has after hours and during vacation (Sonnentag & Fritz, 2007). People usually do not have total control of their free time or vacation time. For example, many workers are forced to stay connected to work during vacation. Another example is when you travel with family members; sometimes you will have to accommodate their needs or preferences.

Autonomy is recognized as a prerequisite for leisure (Linda, 2005). As indicated by the leisure compensation theory (Chick & Hood, 1996), since people usually suffer a loss of autonomy in the workplace, they will tend to choose leisure activities that help to regain their autonomy. The linkage between autonomy and recovery is also suggested by the conservation of resources theory in organizational behavior (Hobfoll, 1989), which maintains that most people will accumulate external resources (such as properties or money) and internal resources (such as positive emotions), while a loss of resources will result in an increase in stress. Therefore, regaining autonomy during non-work time will help a worker to obtain more internal resources, aiding the process of recovery (Sonnentag & Fritz, 2007).

Moreover, workers may also spend their leisure time in gaining passive experiences, which are associated with the recovery mechanisms of relaxation and detachment. Relaxation can be defined as whether an individual will be able to feel relaxed during non-work time or vacation (Sonnentag & Fritz, 2007). Even though relaxation has long been regarded as a key motivational factor for leisure travel, tourism scholars have paid little attention to examine relaxation as a psychological mechanism during leisure travel (Chen et al., 2016). Another similar factor is detachment, which refers to "individual's sense of being away from the work situation" (Etzion, Eden, & Lapidot, 1998, p.579).

As indicated by the effort recovery theory (Meijman & Mulder, 1998), a stressful working environment can bring negative effects to workers, such as negative emotions, fatigue, or loss of energy, while these effects can be reversed if the workers are no longer exposed to the stressful environment. Therefore, if workers are able to feel relaxed or stay detached from work demands, they are more likely to recover from work stress (Newman et al., 2014). The above-mentioned conservation of resources of theory (Hobfoll, 1989) also suggests that being able to feel relaxed and stay detached from work can help workers gain more internal resources required for continued functioning and well-being (Newman et al., 2014).

While workers can choose to passively stay detached from work demands during vacation, they can also actively participate in activities that provide challenging experiences (such as participating in adventure or sports activities) and learning opportunities (such as visiting a museum or a heritage site), which refer to the final mechanism of mastery (Newman et al., 2014). Arguably, having mastery experiences during vacation can help workers obtain internal resources as well as reverse the load reactions caused by work stress (Sonnentag & Fritz, 2007). This connection between mastery experiences and stress relief is also supported by the leisure compensation theory (Chick & Hood, 1996), which maintains that workers will pursue leisure activities that provide satisfaction not realized in the workplace.

Based on the above discussions about the potential role of leisure travel as a stress reliever, a theoretical model was proposed and illustrated in Figure 1. It was proposed that perceived autonomy, relaxation, detachment, and mastery during leisure travel are predictors of perceived life satisfaction after the travel, while the effect of perceived autonomy on life satisfaction is proposed to be moderated by the other three factors. Autonomy is closely related to the concept of perceived freedom in leisure (Newman et al., 2014). Perceived freedom has been recognized as the prerequisite and critical regulator in determining an individual's leisure involvement (Munson, 1993). In other words, when an individual perceives a higher level of freedom, he or she is more likely to control what happens during leisure, while individuals who perceive a lower level of freedom are very likely to miss out opportunities on receiving the benefits of leisure participation (Agyar, 2014). Therefore, the following research hypotheses were formulated:

H1: Perceived autonomy has a positive effect on perceived relaxation during leisure travel.

H2: Perceived autonomy has a positive effect on perceived detachment from work during leisure travel.

H3: Perceived autonomy has a positive effect on perceived mastery experiences during leisure travel.

H4: Perceived relaxation during leisure travel has a positive effect on perceived life satisfaction after the travel.

H5: Perceived detachment from work during leisure travel has a positive effect on perceived life satisfaction after the travel.

H6: Perceived mastery experiences during leisure travel has a positive effect on perceived life satisfaction after the travel.

This research also proposed to examine the moderating effects of trip duration. The effect of vacation length on well-being has sparked scholars' attention, partly because of its importance for policymakers and organizational executives. They can use the findings as a tool to determine public and corporate policies pertaining to holidays and vacations with the aim of improving workers' general well-being, which in turn promotes social welfare and organizational effectiveness.

Despite the practicality of topic and scholars' excitement, research on the relationship between trip duration and well-being remains inconclusive. Kirillova and Wang (2016) found no correlation between vacation duration and vacation recovery. Similarly, Kemp et al.'s (2008) study indicated no relationship between trip duration (ranging from 4 to 14 days) and happiness. However, Etzion (2003) conducted a longitudinal study and found that both long vacations (more than 10 days) and short vacations (7–10 days) helped individuals

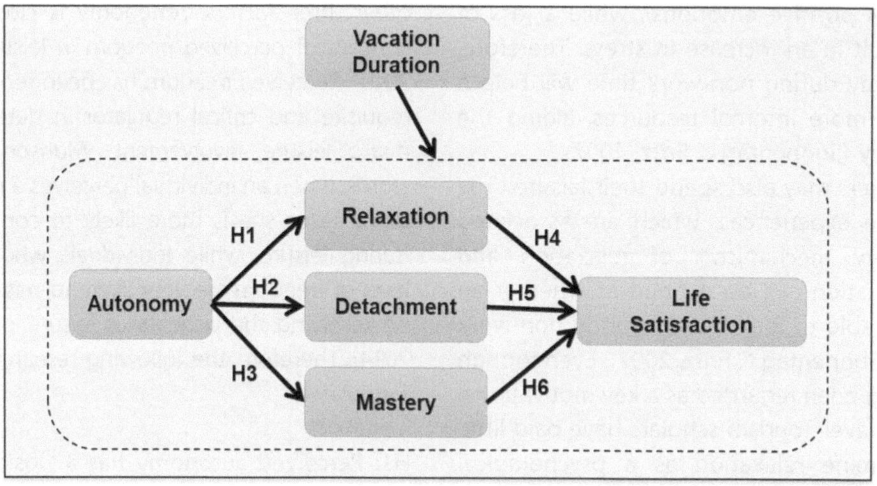

Figure 1. Proposed model.

recover from work demands, but the effect of long vacations was no different from that of short vacations. Additionally, de Bloom, Geurts, and Kompier (2010) reported that the effect of vacation on well-being was similar in moderately long (9 days) and short (4–5 days) vacation. Neal et al. (2007), in sharp contrast, found a moderation effect of trip duration for the relationship between leisure satisfaction and life satisfaction. Specifically, when spending six or fewer nights on vacations, leisure satisfaction lessened life satisfaction, whereas when spending seven or more nights on vacations, leisure satisfaction increased life satisfaction.

Therefore, existing results of effects of trip duration on vacationers' well-being are mixed, and more research to validate the relationship is needed. Given the four recovery mechanisms identified, we propose that shorter and longer leisure trips might facilitate the recovery process in different ways. Previous studies have demonstrated that shorter periods of time away from work such as after hours and weekends tend to aid the psychological mechanisms of relaxation and detachment (Sonnentag & Fritz, 2007). For instance, de Bloom, Geurts, and Kompier (2012) examined the recovery effects of short vacations (4–5 days) and found highly rated relaxation and detachment from work, but no mastery experiences were found.

Methods

Research instruments

This research intended to identify and model the psychological mechanisms that leisure travel experiences can trigger recovery experiences. The survey questionnaire included items measuring five theoretical constructs (autonomy, relaxation, detachment, mastery, and life satisfaction) and a few demographic questions. These questions were used to assess how US tourists' tourism experiences influenced their life satisfaction after their most recent leisure travel.

The four psychological mechanisms of stress relief were measured with a 15-item, 5-point Likert type (1 = strongly disagree to 5 = strongly agree) Recovery Experience Questionnaire (REQ), developed by Sonnentag and Fritz (2007). The REQ includes four items associated with autonomy, detachment, and mastery, and three items associated with relaxation. The scale has frequently been used in the field of organizational behavior (Ragsdale & Beehr, 2016). The concept of life satisfaction was measured using 5-item, 7-point Likert type (1 = strongly disagree to 7 = strongly agree) Satisfaction with Life Scale (SWLS), developed by Diener, Emmons, Larsen, and Griffin (1985). The SWLS was developed to assess satisfaction with people's lives as a whole rather than specific life domains, and has been widely adopted in psychology (Whisman & Judd, 2016) and tourism (McCabe & Johnson, 2013). Respondents were also asked to specify the length and purpose of their most travel, as well as their demographics, including gender, age, marriage, education, income, and employment status.

Data collection

The study population was defined as full-time employees in the United States who were 18 years old or over at the time of data collection. In order to obtain a national representative sample with a limited budget, survey respondents were selected from an online panel company's database, which had more than 30 million panellists in the United States. Nearly 5,000 online panellists were invited to participate in the survey in 2012, while only 1,139 respondents completed the survey, resulting in a response rate of 28.5%. Moreover, a total of 630 respondents were excluded from data analysis because of one of the following reasons: (1) they did not have a leisure trip within three months of data collection, (2) they did not have a full-time job at the time of data collection, or (3) they travelled with children under 18 years old during their most recent leisure travel. It is worth noting that respondents who travelled with their children were excluded from data analysis because travelling with and without children would have a significant impact on travel experiences (Durko & Petrick, 2016). In order to control for the common method bias (e.g. one-time survey administration), we used practices, according to MacKenzie and Podsakoff (2012), such as, explaining and stressing the importance of conscientiousness and accuracy, and encouraging respondents to carefully weigh the alternatives before responding.

Data analysis

Data analysis in this research had six steps. As this research intended to obtain a nationally representative sample, the first step of data analysis involved comparing the demographics of the research sample and the US population using chi-square tests. Subsequently, the skewness and kurtosis values for each item were examined to test the normality assumption required for structural modelling. In the third step, a measurement model with all five constructs was developed using confirmatory factor analysis (CFA) as an attempt to test the fits of the measures. In the next step, a baseline path model was developed using structural equation modelling (SEM) to test H1 to H6. Respondents were further categorized into three groups based on the duration of their most recent leisure travel, including: (1) 1–2 days short vacation (n = 80), (2) 3–7 days medium vacation (n = 312), and (3) 8 days or more long vacation (n = 117). Three path models were then developed and compared using multi-group analysis. Finally, the total effects of autonomy on life satisfaction in three models were statistically compared using bootstrapping.

Results

Respondents' profile

The profile of respondents was first examined. As shown in Table 2, nearly equal numbers of female (n = 242; 47.5%) and male respondents (n = 267; 52.5%) completed the survey. According to the latest US census data in 2010 (US Census Bureau, 2011), females and males represented

Table 2. Profile of respondents.

Variables	n	%	US Population %[a]
Gender[b]			
Male	267	52.5	49.2
Female	242	47.5	50.8
Marital Status[c]			
Married	196	39.4	51.1
Single (including divorced, separated, and widowed)	302	60.6	48.9
Age (years)[d]			
Under 20	13	2.6	
20–29	103	20.2	18.9
30–39	102	20.0	17.8
40–49	90	17.7	19.3
50–59	122	24.0	18.6
60–69	47	9.2	13.0
70 and over	32	6.3	12.3
Education[e]			
High School	100	19.6	44.1
College	271	53.2	45.3
Graduate School	138	27.1	10.5
Income (US$)[f]			
Under $25,000	107	21.0	17.8
$25,000 – 49,999	142	27.9	23.8
$50,000 – 74,999	132	25.9	19.4
$75,000 – 99,999	48	9.4	13.5
$100,000 and above	80	15.7	25.6

[a]The percentages for the US population were drawn from the 2010 Census Data.
[b]Result of chi-square test: χ^2 = 2.159; df = 1; p > .05
[c]Result of chi-square test: χ^2 = 27.481; df = 1; p < .001
[d]Result of chi-square test: χ^2 = 32.018; df = 5; p < .001
[e]Result of chi-square test: χ^2 = 212.104; df = 2; p < .001
[f]Result of chi-square test: χ^2 = 43.217; df = 4; p < .001

50.8% and 49.2% of the US population. The result of chi-square test showed that the survey sample and the US population were homogeneous in gender (χ^2 = 2.159; df = 1; p > .05). Regarding respondents' marital status, a majority of them were single (n = 302; 60.6%) and nearly 40% of respondents were married at the time of data collection (n = 196; 39.4%). However, the census data showed that more than half of US population were married in 2010 (51.1%). Therefore, the result of chi-square test indicated that the sample and population were not homogeneous (χ^2 = 27.481; df = 1; p < .001).

Table 2 also showed that a majority of respondents were aged between 20 and 59 (81.9%), including 103 respondents aged 20–29 (20.2%), 102 respondents aged 30–39 (20.0%), 90 respondents aged 40–49 (17.7%), and 122 respondents aged 50–59 (24.0%), while the percentage of US population who aged between 20 and 59 was lower in 2010 (74.6%). Moreover, only a total of 79 respondents (15.5%) were 60 years old or over, while 25.3% of the US population in 2010 were aged 60 or over. Therefore, the result of chi-square showed the sample and population were not homogenous in age (χ^2 = 32.018; df = 5; p < .001).

Regarding respondents' education, a majority of them (79.3%) received undergraduate (n = 271; 53.2%) or graduate education (n = 138; 27.1%), while only 45.3% and 10.5% of the US population had undergraduate and graduate education. The difference between the sample and population in education was significant as indicated by the result of chi-square test (χ^2 = 212.104; df = 2; $p < .001$). The difference between the sample and population in income was further examined. As shown in Table 2, high percentages of respondents had a household income of $25,000 or under (sample: 21.0%; population: 17.8%), $25,000–49,999 (sample: 27.9%; population: 23.8%), and $50,000–74,999 (sample: 25.9%; population: 19.4%). However, lower percentages of respondents had a household income of $75,000–99,999 (sample: 9.4%; population: 13.5%) or $100,000 or above (sample: 15.7%; population: 25.6%). Therefore, the result of chi-square test showed that the difference in income was significant (χ^2 = 43.217; df = 4; $p < .001$).

In summary, the survey sample and the US population were homogenous in gender, while their differences in marital status, age, education, and income were significant ($p < .001$). As compared to the population, more survey respondents were single, and in general survey respondents were younger and more educated with lower income levels.

Table 3. Descriptive statistics of survey items.

Constructs/Items	Mean	S.D.	Kurtosis	Skewness
Autonomy[a]				
*I took care of things the way that I wanted them done.	3.87	.98	.20	–.71
*I determined for myself how I would spend my time.	3.87	1.04	.20	–.84
*I decided my own schedule.	3.87	.99	–.14	–.68
*I felt like I could decide for myself what to do.	3.85	.93	.16	–.60
Detachment				
*I got a break from the demands of work.	3.78	1.09	.11	–.80
*I distanced myself from the demands of work.	3.61	1.17	–.73	–.48
*I didn't think about work at all.	3.98	.98	.31	–.83
*I forgot about work.	4.14	.90	.96	–1.02
Mastery				
*I did things to broaden my horizons.	3.56	1.05	–.37	–.35
*I did things that challenged me.	3.00	1.17	–.72	–.02
*I sought out intellectual challenges.	3.12	1.14	–.67	–.13
*I learned new things.	3.31	1.13	–.56	–.26
Relaxation				
*I took time for leisure.	4.17	.89	1.02	–1.06
*I used the time to relax.	4.11	.91	.39	–.92
*I did relaxing things.	4.14	.91	.96	–.96
Life satisfaction[b]				
*In most ways my life was close to my ideal.	4.44	1.56	–.41	–.35
*The conditions of my life were excellent.	4.61	1.51	–.38	–.40
*I was satisfied with my life.	4.87	1.51	–.25	–.56
*I felt I had the important things I wanted in life.	4.97	1.45	–.12	–.55
*If I could have lived my life over, I would change almost nothing.	4.30	1.69	–.75	–.33

[a]Autonomy, detachment, mastery, and relaxation were measured using 5-point scales.
[b]Life satisfaction was measured with a 7-point scale.

Descriptive statistics

All 20 survey items are listed in Table 3, including 15 5-point items measuring the four psychological mechanisms of stress relief and 5 7-point items of life satisfaction. Regarding the four psychological mechanisms of stress relief, items associated with relaxation had higher scores, followed by items measuring autonomy. However, items associated with mastery had lower scores and greater standard deviations.

The kurtosis and skewness values of each item were also examined in order to test the normality assumption for structural modelling. According to Hair, Anderson, Tatham, and Black (1998), kurtosis and skewness values between −1 and 1 indicate uni-variate normality, while the absolute value of the kurtosis greater than 8 and the absolute value of the skewness greater than 3 should be considered as significant departure from normality (Kline, 2010). As shown in Table 3, all kurtosis and skewness values fell within the range of −1 and 1, except the skewness value for the item *I forgot about work* (−1.02) and the kurtosis (1.02) and skewness values (−1.06) for the item *I took time for leisure*. As these three values were very close to 1 or −1, the normality assumption was met.

Measurement model

Subsequently, the fits of measures were assessed using CFA. Based on Byrne's (1998) and Bollen's (1989) suggestions, a model is regarded as acceptable if the chi-square to df ratio is lower than 3, the standardize root-mean-square residual (SRMR) is smaller than 0.08, the comparative fit index (CFI) exceeds .93, the normed fit index (NFI) and, and the root-mean-square error of approximation (RMSEA) is less than .80. Results from CFA showed that the initial measurement model, consisting of three 4-item factors (autonomy, detachment, and mastery), one 3-item factor (relaxation), and one 5-item factor (life satisfaction) had poor fit indices ($\chi 2$ = 641.10, df = 160, $\chi 2$/df = 4.01; SRMR = 0.085; CFI = .94, NFI = .92, and RMSEA = .077). Since all 20 regression weights were significant ($p < .001$), the measurement model was refined by deleting items associated with greater residuals and modification indices as suggested by Netemeyer, Bearden, and Sharma (2003). As a result, three items were deleted. Each of them was associated with detachment (I forgot about work), mastery (I did things that challenged me), and life satisfaction (I felt I had the important things I wanted in life). The resulting measurement model had good fit indices ($\chi 2$ = 313.21, df = 109, $\chi 2$/df = 2.87; SRMR = 0.046; CFI = .97, NFI = .95, GFI = .93, and RMSEA = .061).

As shown in Table 4, all 17 items had factor loadings higher than .67, which is considered as a

Table 4. Results of confirmatory factor analysis.

Constructs/Items[a]	Factor loadings[b]	Error variances	CR	AVE
Autonomy[a]			.92	.74
*I took care of things the way that I wanted them done.	.82	.32		
*I determined for myself how I would spend my time.	.91	.17		
*I decided my own schedule.	.92	.15		
*I felt like I could decide for myself what to do.	.79	.38		
Detachment			.90	.75
*I got a break from the demands of work.	.92	.15		
*I distanced myself from the demands of work.	.85	.27		
*I didn't think about work at all.	.82	.33		
Mastery			.86	.67
*I did things to broaden my horizons.	.77	.41		
*I sought out intellectual challenges.	.78	.39		
*I learned new things.	.90	.19		
Relaxation			.91	.78
*I took time for leisure.	.88	.22		
*I used the time to relax.	.90	.19		
*I did relaxing things.	.86	.26		
Life satisfaction[b]			.92	.73
*In most ways my life was close to my ideal.	.87	.24		
*The conditions of my life were excellent.	.96	.07		
*I was satisfied with my life.	.90	.20		
*If I could have lived my life over, I would change almost nothing.	.67	.55		

[a]Fit indices $\chi 2 = 313.21$, df = 109, $\chi 2/df = 2.87$; SRMR = 0.046; CFI = .97, NFI = .95, GFI = .93, and RMSEA = .061
[b]All factor loadings were significant at the .0001 level.

high factor loading as suggested by Guadagnoli and Velicer (1988). The convergent validity of the measures was further examined using composite reliability (CR) and average variance extracted estimate (AVE). The CR values for all five constructs were higher than .86, which exceeded the suggested value of .80 (Netemeyer et al., 2003). Similarly, the AVEs of all five constructs were greater than .67, which was also higher than the suggested value of .50 (Netemeyer et al., 2003). These findings suggested that high convergent validity of the measures.

The discriminant validity of the measures was further examined by comparing the AVEs of two constructs and the square of the correlation coefficient between these two constructs. The results of CFA showed that all ten correlation coefficients were significant ($p < .001$), and the values ranged from .21 to .62. As the squares of the correlations coefficients were much than lower the AVEs of five constructs, the discriminant validity of the measures was considered as high.

Baseline model

In the next step, H1 to H6 were tested by establishing a baseline path model (n = 509). As shown in Figure 2, the baseline path model had acceptable fit indices ($\chi 2 = 372.73$, $df = 113$, $\chi 2/df = 3.30$; SRMR = 0.071; CFI = .96, NFI = .95, and RMSEA = .068). Results from SEM revealed that all six proposed effects were positive and significant ($p < .001$), including the effects of autonomy on detachment ($\beta = .45$; $p < .001$), relaxation ($\beta = .65$; $p < .001$), and mastery ($\beta = .38$; $p < .001$), as well as the effects of detachment ($\beta = .23$; $p < .001$), relaxation ($\beta = .17$; $p < .001$), and mastery ($\beta = .25$; $p < .001$) on life satisfaction. These findings suggest that H1 to H6 were all supported.

Moreover, the standardized total effect of autonomy on life satisfaction was .31, which suggests autonomy explained approximately 10% of the variance associated with life satisfaction. As proposed in this research, autonomy affects life satisfaction through the passive route (relaxation and detachment) and the active route (mastery). It was found that autonomy had greater effects on life satisfaction through the passive route (relaxation: $\beta = .11$; detachment: $\beta = .10$) than the active route (mastery: $\beta = .10$).

Multi-group analysis

The next step of data analysis involved categorizing all respondents into three groups based on the duration of their most recent leisure travel. These three groups included: (1) short-vacation group (n = 80) with respondents who travelled for only 1–2 days, (2) medium-vacation group (n = 312) with respondents who travelled for 3–7 days, and (3) long-vacation group (n = 117) with respondents who travelled for 8 days or more.

As shown in Figure 2, three separate path models were established. Overall, the structural weights among the three models appeared to be different. Regarding the short-vacation model, only four of the six proposed effects were significant ($p < .05$), including: the effects of autonomy on relaxation ($\beta = .77$; $p < .001$) and mastery ($\beta = .26$; $p < .05$) and the effects of relaxation ($\beta = .37$; $p < .01$) and mastery ($\beta = .31$; $p < .01$) on life satisfaction, while the effects of autonomy on detachment ($\beta = .19$; $p > .05$) and the effect of detachment on life satisfaction ($\beta = .10$; $p > .05$) were not significant. The standardized total effect of autonomy on life satisfaction was .39, while it was found that the mediating effect of relaxation ($\beta = .29$) was greater than the mediating effects of detachment ($\beta = .02$) and mastery ($\beta = .08$).

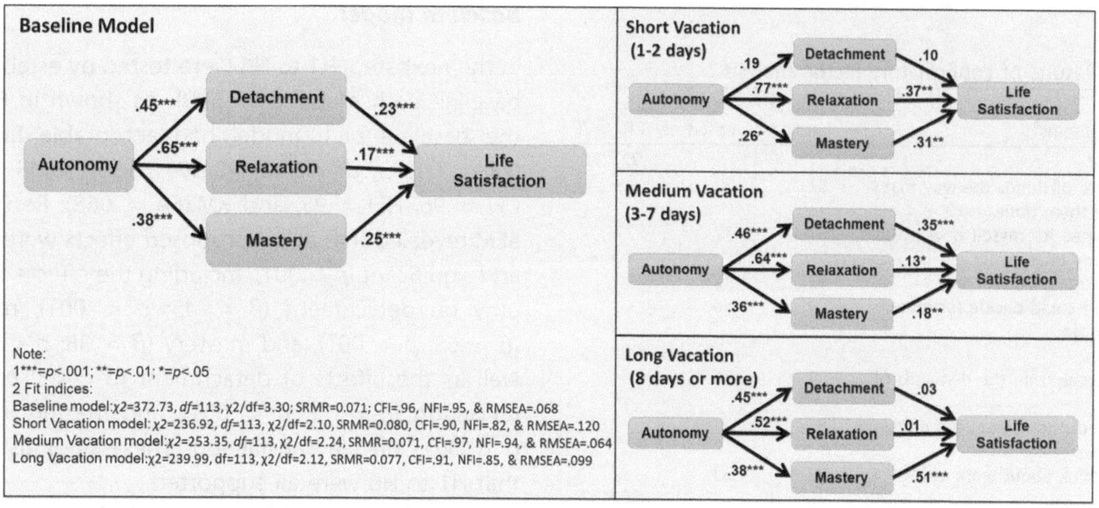

Figure 2. Results of structural modelling.

The medium-vacation model was subsequently examined. Results from SEM showed that all six proposed effects were significant ($p < .05$). The standardized total effect of autonomy on life satisfaction was .31, while the mediating effect of detachment ($\beta = .16$) was greater than the mediating effects of relaxation ($\beta = .08$) and mastery ($\beta = .06$). Regarding the long-vacation model, only four of the six proposed effects were significant ($p < .001$), including: the effects of autonomy on detachment ($\beta = .45$; $p < .001$), relaxation ($\beta = .52$; $p < .001$) and mastery ($\beta = .38$; $p < .001$) and the effects of mastery on life satisfaction ($\beta = .51$; $p < .001$), while the effects of detachment ($\beta = .03$; $p > .05$) and relaxation ($\beta = .01$; $p > .05$) on life satisfaction were not significant. The standardized total effect of autonomy on life satisfaction was .21, while it was found that the mediating effect of mastery ($\beta = .20$) was greater than the mediating effects of detachment ($\beta = .01$) and relaxation ($\beta = .01$).

Finally, the differences among the three models were statistically tested using multi-group analysis, which involved three pairs of comparison. The differences between the short-vacation and medium-vacation models were first compared. As shown in Table 5,

Table 5. Results of multi-group analysis.

Models	χ2	*df*	Δχ2	Δ*df*	*p*	Note
Short Vacation vs. Medium Vacation						
Unconstrained	491.58	226				
Structural weights	509.39	232	17.81	6	<.01	
Autonomy → Detachment	496.60	227	5.11	1	<.05	Medium > Short Vacation
Autonomy → Relaxation	494.63	227	3.05	1	>.05	
Autonomy → Mastery	492.24	227	0.66	1	>.05	
Detachment → Life satisfaction	494.28	227	2.70	1	>.05	
Relaxation → Life satisfaction	494.05	227	2.47	1	>.05	
Mastery → Life satisfaction	492.85	227	1.27	1	>.05	
Short Vacation vs. Long Vacation						
Unconstrained	477.08	226				
Structural weights	495.64	232	18.56	6	<.01	
Autonomy → Detachment	480.62	227	3.54	1	>.05	
Autonomy → Relaxation	483.56	227	6.48	1	<.01	Short > Long Vacation
Autonomy → Mastery	477.73	227	0.65	1	>.05	
Detachment → Life satisfaction	477.38	227	0.30	1	>.05	
Relaxation → Life satisfaction	481.37	227	4.29	1	<.05	Short > Long Vacation
Mastery → Life satisfaction	478.23	227	1.15	1	>.05	
Medium Vacation vs. Long Vacation						
Unconstrained	493.90	226				
Structural weights	511.23	232	17.33	6	<.01	
Autonomy → Detachment	493.91	227	0.01	1	>.05	
Autonomy → Relaxation	495.91	227	2.01	1	>.05	
Autonomy → Mastery	493.93	227	0.03	1	>.05	
Detachment → Life satisfaction	501.00	227	7.10	1	<.01	Medium > Long Vacation
Relaxation → Life satisfaction	494.83	227	0.93	1	>.05	
Mastery → Life satisfaction	501.29	227	7.39	1	<.01	Long > Medium Vacation

the results of multi-group analysis showed that the unconstrained model had a chi-square value of 491.58 with 226 degrees of freedom, while the chi-square value increased to 509.39 by constraining all six paths to be equal, which suggested that the short-vacation and medium-vacation models were significantly different in structural weights ($\Delta\chi 2 = 17.81$; $\Delta df = 6$; $p < .01$). The differences between the two models in each structural weight were first examined. It was found that the effect of autonomy on detachment in the medium-vacation model was significantly higher than the effect in the short-vacation model ($\Delta\chi 2 = 5.11$; $\Delta df = 1$; $p < .05$), while the differences between two models in the other five structural weights were not found to be significant ($p > .05$).

The second pair of comparison involved the short-vacation and long-vacation models. As shown in Table 5, the two models were significantly different across six structural weights ($\Delta\chi 2 = 18.56$; $\Delta df = 6$; $p < .01$). The subsequent analysis revealed that the effect of autonomy on relaxation in the short-trip model was significantly higher than the effect in the long-vacation model ($\Delta\chi 2 = 6.48$; $\Delta df = 1$; $p < .01$). It was also found that the effect of relaxation on life satisfaction in the short-vacation model was significantly higher than the effect in the long-vacation model ($\Delta\chi 2 = 4.29$; $\Delta df = 1$; $p < .05$), while the differences in the other four structural weights were not found to be significant ($p > .05$).

Lastly, the differences between the medium-vacation and long-vacation models. The results of multi-group analysis revealed that the two modes differed significantly across six structural weights ($\Delta\chi 2 = 17.35$; $\Delta df = 6$; $p < .01$). Regarding the difference in each structural weight, it was found that the effect of detachment on life satisfaction in the medium-vacation model was significantly greater the effect in the long-vacation model ($\Delta\chi 2 = 7.10$; $\Delta df = 1$; $p < .01$), while the effect of mastery on life satisfaction in the long-vacation model was greater than the effect in the medium-vacation model ($\Delta\chi 2 = 7.39$; $\Delta df = 1$; $p < .01$). The differences in the other five structural weights were not found to be significant ($p > .05$).

In summary, the three models differed significantly in structural weights ($p < .01$), which suggests that tourism experiences can relieve tourists' stress through different routes. In particular, it was found that tourism experiences tend to affect life satisfaction through the route (autonomy → relaxation or detachment → life satisfaction) for shorter leisure trips (7 days or less), while for longer leisure trips (8 days or more), tourism experiences tend to affect life satisfaction through the active route (autonomy → mastery → life satisfaction). Another interesting finding is that the effect of autonomy on life satisfaction appeared to diminish with the length of leisure travel. Moreover, the results of bootstrapping further revealed that the total effects of autonomy on life satisfaction in all three models were significant at the .01 level. It was also found that the 95% confidence intervals for the total effects were between .18 and .57 (short vacation), between .21 and .41 (medium vacation), and between .06 and .37 (long vacation). As the three intervals were overlapped, the differences of the total effects of autonomy on life satisfaction in three models were not significant ($p > .05$).

Conclusions and implications

Given that an increasing number of workers across the world are experiencing stress in the workplace, this research examined how leisure travel can help recovery from work. Derived from a sample of 509 US respondents, the study results showed that autonomy is a requisite in the link between tourism experiences and life satisfaction, while its effect on life satisfaction is mediated by relaxation, detachment, and mastery. It was also found that shorter leisure trips tend to facilitate the passive route to life satisfaction (detachment and relaxation), while longer trips can better facilitate the active route (mastery). These findings have important theoretical and practical implications.

Theoretically, the topic of tourism and well-being has attracted increasing attention in the tourism literature (Uysal et al., 2016), and previous research on this topic has demonstrated the connection between tourism experience and well-being (Chen & Petrick, 2013). This study further identified the recovery experiences associated with vacations as important contributors to life satisfaction. Arguably, these findings not only confirm the positive effect of tourism experiences on tourists' subjective well-being but also uncovered *how* tourism experiences contribute to workers' well-being by triggering the psychological mechanisms of recovery.

Building on Sonnentag and Fritz's (2007) conceptualization of recovery experience, this research extends previous work on recovery experiences by examining how the four psychological mechanisms of recovery are interrelated. Specifically, the current study developed and tested a model of vacation recovery process, in which autonomy is proposed and verified as an antecedent of relaxation, detachment and mastery experiences. The findings imply the essential role of autonomy in leisure as well as vacation experience, which is in line with leisure scholars' view that autonomy is a necessary requisite of leisure (Newman et al., 2014).

Furthermore, as indicated by the study results, while regaining autonomy plays a key role in the process of recovery during leisure travel, the process of recovery can also be facilitated for workers to use their free time to simply stay away from work demands (detachment and relaxation experiences) or actively seek out challenging or learning experiences (mastery experiences) (de Bloom, Geurts, & Kompier, 2013). These findings help to fill the gap in the literature because previous studies have tended to examine "whether" tourism experiences and well-being have a connection, but these findings reveal "how" the connection is established. Therefore, this research adds knowledge to the topic of tourism experience and well-being.

This research further examined the moderating effects of trip duration. The study results suggest that workers may not benefit more from taking a longer leisure travel, while longer trips can better facilitate the active recovery experiences (mastery experiences) and shorter trips tend to aid the passive recovery experiences (detachment and relaxation experiences). These findings echo de Bloom et al.'s (2012) study which found that in medium vacations (4–5 days), mastery experiences hardly occurred, but relaxation and detachment from work were detected. Arguably, these findings provide important details regarding the relationships among tourism experiences, recovery, and well-being because vacation plans are often constrained by the availability of time and money (Chen & Petrick, 2016). As demonstrated in this research, having longer and shorted vacation days can bring different values for workers.

The findings of this study also provide important practical implications for the tourism industry and government policymakers. For the tourism industry, it is important to help their customers achieve a successful recovery from work during their stays and visits not only for the increased well-being purpose, but also because evidence suggests that these efforts will lead to customer satisfaction (Chen et al., 2016; Uysal et al., 2016). Specific strategies include providing different activities for each family member to facilitate the mechanism of autonomy or encouraging tourists to choose staying unplugged from work during their stays to aid the mechanism of detachment (Chen et al., 2016). It is worth noting that tourists might be encouraged to use smartphone for the non-work purpose, such as using social media to post-vacation pictures, that could enrich the tourist experience, according to Kirillova and Wang (2016).

Another strategy might be providing educational opportunities for tourists to learn local culture and cuisine, which will help tourist gain recovery experiences associated with mastery. Although many hotels and tourism service providers have used these strategies by simply focusing on the hedonic benefits of tourism experiences, it is more important to communicate with tourists about the recovery mechanisms of tourism experiences that might help better recover from work stress (Chen & Petrick, 2016). For example, tourism professionals should include the information of recovery in their promotion booklets and advertisement brochure to educate tourists about the importance and practices of recovery during vacation. More practically, hotel and resorts might also want to design programs and provide experiences to facilitate the recovery process for tourists. For example, they can offer a destination filming program or one-time free yoga class for their consumers and emphasize the recovery information while on vacation.

Moreover, the study findings suggest that longer vacations are more likely to facilitate mastery experience whereas shorter vacations are more likely to result in detachment and relaxation experiences. On one hand, tourism professionals can educate tourists on the best practices of planning the length of their vacations. On the other hand, depending upon tourists' vacation motivation, tourism professionals should recommend longer vacations (8 days or more) to tourists whose primary vacation motivation is to seek challenges and promote shorter vacations (fewer than 7 days) to those who are looking for relaxation and escape from vacations.

Study results are also insightful and meaningful for employers and government policymakers. As demonstrated in this research, taking leisure trips is beneficial for workers, and longer and shorter leisure trips can facilitate the process of recovery in different ways. Since high work stress can lead to major health problems for workers (Wilkins & Beaudet, 1998) and lower organizational performance (Briner & Reynolds, 1999), it is important to provide workers opportunities to take vacations. When a vacation includes many active and challenging elements (e.g. seeking out intellectual challenges), the vacation is suggested to be 8 days or long; however, when a vacation is considered passive (e.g. beach time), it should be planned within 7 days preferably some vacations with seven days or more. Moreover, as discussed earlier, workers in North America and East Asia typically work for longer hours, which seems to be associated with lower life satisfaction in these regions (Organization for Economic Cooperation and Development, 2017). Policymakers and governments in these countries, thus, should consider providing incentives for companies and organizations in order to provide their employees more paid vacations days.

Finally, it is worth mentioning the limitations of the study and proposing recommendations for future research. First, while this research examined how leisure travel can help full-time workers to recover from vacations, this research did not measure job-related variables (e.g. respondents' occupations, job titles), which future research should consider including for more in-depth analyses. Moreover, this research examined tourism experiences as a factor influencing life satisfaction, while other determiners of life satisfaction were not measured, including personality or satisfaction with specific life domains. In addition, this study only measured life satisfaction as one dimension of subjective well-being, while other dimensions were not included, such as positive affect and negative affect (Diener, 1984). It is important that future research considers measuring additional dimensions of well-being to gain more insights. Another conceptual-related limitation is about defining active and passive vacation experiences. Although this study made efforts to distinguish between these two concepts, future researchers should continue examining the inclusive elements within these two concepts in order to better clarify active and passive experience in the vacation context.

Disclosure statement

No potential conflict of interest was reported by the authors.

ORCID

Suiwen (Sharon) Zou http://orcid.org/0000-0002-9692-240X

References

Agyar, E. (2014). Contribution of perceived freedom and leisure satisfaction to life satisfaction in a sample of Turkish women. *Social Indicators Research*, *116*(1), 1–15.

Belleau, C., DuPont, R., Erickson, C., Flaherty, M., Galanter, M., Gold, M., ... White, W. (2007). What is recovery? A working definition from the Betty Ford Institute. *Journal of Substance Abuse Treatment*, *33*(3), 221–228.

Bollen, K. A. (1989). *Structural equations with latent variables*. New York: Wiley.

Briner, R. B., & Reynolds, S. (1999). The costs, benefits, and limitations of organizational level stress interventions. *Journal of Organizational Behavior*, *20*(5), 647–664.

Byrne, B. M. (1998). *Structural equation modeling with LISREL, PRELIS, and SIMPLIS: Basic concepts, applications, and programming*. Mahwah, NJ: Lawrence Erlbaum Associates.

Caldwell, L. L. (2005). Leisure and health: Why is leisure therapeutic? *British Journal of Guidance & Counselling*, *33*(1), 7–26.

Chen, C. C., & Petrick, J. F. (2013). Health and wellness benefits of travel experiences: A literature review. *Journal of Travel Research*, *52*(6), 709–719.

Chen, C. C., Huang, W. J., & Petrick, J. F. (2016). Holiday recovery experiences, tourism satisfaction and life satisfaction– Is there a relationship? *Tourism Management*, *53*, 140–147.

Chen, C. C., Petrick, J. F., & Shahvali, M. (2016). Tourism experiences as a stress reliever: Examining the effects of tourism recovery experiences on life satisfaction. *Journal of Travel Research*, *55*(2), 150–160.

Chen, C. C., & Petrick, J. F. (2016). The roles of perceived travel benefits, importance, and constraints in predicting travel behavior. *Journal of Travel Research*, *55*(4), 509–522.

Chen, C. C., Huang, W. J., Gao, J., & Petrick, J. F. (2017). Antecedents and consequences of work-related smartphone use on vacation: An exploratory study of Taiwanese tourists. *Journal of Travel Research*, *57(6)*, 734-756.

Chick, G., & Hood, R. D. (1996). Working and recreating with machines: Outdoor recreation choices among machine-tool workers in western Pennsylvania. *Leisure Sciences*, *18*(4), 333–354.

Combs, J., Liu, Y., Hall, A., & Ketchen, D. (2006). How much do high-performance work practices matter? A meta-analysis of their effects on organizational performance. *Personnel Psychology*, *59*(3), 501–528.

de Bloom, J., Geurts, S., & Kompier, M. (2010). Vacation from work as prototypical recovery opportunity. *Gedrag & Organisatie*, *23*, 333–349.

de Bloom, J., Geurts, S., & Kompier, M. (2012). Effects of short vacations, vacation activities and experiences on employee health and well-being. *Stress and Health*, *28*(4), 305–318.

de Bloom, J., Geurts, S., & Kompier, M. (2013). Vacation (after-) effects on employee health and well-being, and the role of vacation activities, experiences and sleep. *Journal of Happiness Studies*, *14*, 613–633.

de Bloom, J., Kompier, M., Geurts, S., de Weerth, C., Taris, T., & Sonnentag, S. (2009). Do we recover from vacation? Meta-analysis of vacation effects on health and well-being. *Journal of Occupational Health*, *51*, 13–25.

Diener, E. (1984). Subjective well-being. *Psychological Bulletin*, *95*(3), 542–575.

Diener, E. D., Emmons, R. A., Larsen, R. J., & Griffin, S. (1985). The satisfaction with life scale. *Journal of Personality Assessment*, *49*(1), 71–75.

Dolnicar, S., Yanamandram, V., & Cliff, K. (2012). The contribution of vacations to quality of life. *Annals of Tourism Research*, *39*(1), 59–83.

Durko, A. M., & Petrick, J. F. (2016). Travel as relationship therapy: Examining the effect of vacation satisfaction applied to the investment model. *Journal of Travel Research*, *55*(7), 904–918.

Etzion, D. (2003). Annual vacation: Duration of relief from job stressors and burnout. *Anxiety, Stress, and Coping*, *16*(2), 213–226.

Etzion, D., Eden, D., & Lapidot, Y. (1998). Relief from job stressors and burnout: Reserve service as a respite. *Journal of Applied Psychology*, *83*(4), 577–585.

Expedia. (2017). *2016 Expedia vacation deprivation study*. Retrieved July 1, 2015, from http://blog.expedia.co.uk/vacation-deprivation-2016/

Fritz, C., & Sonnentag, S. (2006). Recovery, well-being, and performance-related outcomes: The role of workload and vacation experiences. *Journal of Applied Psychology*, *91*(4), 936–945.

Gilbert, D., & Abdullah, J. (2004). Holidaytaking and the sense of well-being. *Annals of Tourism Research*, *31*(1), 103–121.

Guadagnoli, E., & Velicer, W. (1988). Relation of sample size to the stability of component patterns. *Psychological Bulletin*, *103*, 265–275.

Hair, J. F., Anderson, R. E., Tatham, R. L., & Black, W. C. (1998). *Multivariate data analysis*. New Jersey: Prentice-Hall.

Hobfoll, S. E. (1989). Conservation of resources: A new attempt at conceptualizing stress. *American Psychologist*, *44*(3), 513–524.

Iwasaki, Y., Zuzanek, J., & Mannell, R. C. (2001). The effects of physically active leisure on stress-health relationships. *Canadian Journal of Public Health*, *92*(3), 214–218.

Kemp, S., Burt, C. D. B., & Furneaux, L. (2008). A test of the peak-end rule with extended autobiographical events. *Memory and Cognition*, *36*, 502–514.

Kim, E., Chiang, L., & Tang, L. (2017). Investigating wellness tourists' motivation, engagement, and loyalty: In search of the missing link. *Journal of Travel & Tourism Marketing*, *34* (7), 867–879.

Kim, H. J., Lee, T. J., & Ko, T. G. (2016). Satisfaction and subjective well-being of health tourists: The case of Japanese and Korean tourists. *Journal of Travel & Tourism Marketing*, *33*(5), 742–756.

Kinnunen, U., Rantanen, J., de Bloom, J., Mauno, S., Feldt, T., & Korpela, K. (2016). The role of work–Nonwork boundary management in work stress recovery. *International Journal of Stress Management*, *23*(2), 99.

Kirillova, K., & Wang, D. (2016). Smartphone (dis)connectedness and vacation recovery. *Annals of Tourism Research*, *61*, 157–169.

Kline, R. B. (2010). *Principles and practice of structural equation modeling*. New York, NY: Guilford Press.

Kühnel, J., & Sonnentag, S. (2011). How long do you benefit from vacation? A closer look at the fade-out of vacation effects. *Journal of Organizational Behavior*, *32*(1), 125–143.

Lehto, X. (2013). Assessing the perceived restorative qualities of vacation destinations. *Journal of Travel Research*, *52*(3), 325–339.

McCabe, S., & Johnson, S. (2013). The happiness factor in tourism: Subjective well-being and social tourism. *Annals of Tourism Research*, *41*, 42–65.

McCabe, S., Joldersma, T., & Li, C. (2010). Understanding the benefits of social tourism: Linking participation to subjective well-being and quality of life. *International Journal of Tourism Research*, *12*(6), 761–773.

Meijman, T. F., & Mulder, G. (1998). Psychological aspects of workload. In P. J. D. Drenth, H. Thierry, & C. J. De Wolff (Eds.), *Handbook of work and organizational psychology* (Vol. 2: Work Psychology, pp. 5–33) . Hove, England: Psychology Press.

Morgan, N., Pritchard, A., & Sedgley, D. (2015). Social tourism and well-being in later life. *Annals of Tourism Research*, *52*, 1–15.

Munson, W. W. (1993). Perceived freedom in leisure and career salience in adolescence. *Journal of Leisure Research*, *25*(3), 305–314.

Nawijn, J., Marchand, M., Veenhoven, R., & Vingerhoets, A. (2010). Vacationers happier, but most not happier after a holiday. *Applied Research in Quality of Life*, *5*(1), 35–47.

Nawijn, J., Mitas, O., Lin, Y., & Kerstetter, D. (2013). How do we feel on vacation? A closer look at how emotions change over the course of a trip. *Journal of Travel Research*, *52*(2), 265–274.

Neal, J. D., Uysal, M., & Sirgy, M. J. (2007). The effect of tourism services on travelers' quality of life. *Journal of Travel Research*, *46*(2), 154–163.

Netemeyer, R. G., Bearden, W. O., & Sharma, S. (2003). *Scaling procedures: Issues and applications*. Thousand Oaks, CA: Sage Publications.

Newman, D. B., Tay, L., & Diener, E. (2014). Leisure and subjective well-being: A model of psychological mechanisms as mediating factors. *Journal of Happiness Studies*, *15*(3), 555–578.

Organization for Economic Co-operation and Development. (2017). *OECD Statistics*. Retrieved July 15, 2017, from http://stats.oecd.org/

Podsakoff, P. M., MacKenzie, S. B., & Podsakoff, N. P. (2012). Sources of method bias in social science research and recommendations on how to control it. *Annual Review of Psychology*, 63, 539–569.

Ragsdale, J. M., & Beehr, T. A. (2016). A rigorous test of a model of employees' resource recovery mechanisms during a weekend. *Journal of Organizational Behavior*, *37*(6), 911–932.

Sirgy, M. J. (2010). Toward a quality-of-life theory of leisure travel satisfaction. *Journal of Travel Research*, *49*(2), 246–260.

Sirgy, M. J., Kruger, P. S., Lee, D. J., & Yu, G. B. (2011). How does a travel trip affect tourists' life satisfaction? *Journal of Travel Research*, *50*(3), 261–275.

Sonnentag, S., & Fritz, C. (2007). The recovery experience questionnaire: Development and validation of a measure for assessing recuperation and unwinding from work. *Journal of Occupational Health Psychology*, *12*(3), 204–221.

Sonnentag, S., & Fritz, C. (2015). Recovery from job stress: The stressor-detachment model as an integrative framework. *Journal of Organizational Behavior*, *36*(1), 72–103.

Tarumi, K., Hagihara, A., & Morimoto, K. (1998). An investigation into the effects of vacations on the health status in male white-collar workers. *Environmental Health and Preventive Medicine*, *3*(1), 23.

Tennant, C. (2001). Work-related stress and depressive disorders. *Journal of Psychosomatic Research*, *51*(5), 697–704.

Trenberth, L., & Dewe, P. (2002). The importance of leisure as a means of coping with work related stress: An exploratory study. *Counselling Psychology Quarterly*, *15*(1), 59–72.

US Census Bureau. (2011). 2010 US Census Date. Retrieved September 30, 2017, from https://www.census.gov/programs-surveys/decennial-census/decade.2010.html

Uysal, M., Sirgy, M. J., Woo, E., & Kim, H. L. (2016). Quality of life (QOL) and well-being research in tourism. *Tourism Management*, *53*, 244–261.

Wei, S., & Milman, A. (2002). The impact of participation in activities while on vacation on seniors' psychological well-being: A path model application. *Journal of Hospitality & Tourism Research*, *26*(2), 175–185.

Westman, M., & Eden, D. (1997). Effects of a respite from work on burnout: Vacation relief and fade-out. *Journal of Applied Psychology*, *82*(4), 516.

Whisman, M. A., & Judd, C. M. (2016). A cross-national analysis of measurement invariance of the satisfaction with life scale. *Psychological Assessment*, *28*(2), 239–244.

Wilkins, K., & Beaudet, M. P. (1998). Work stress and health. *Health Report*, *10*(3), 47–62.

Listening to the murmur of water: essential satisfaction and dissatisfaction attributes of thermal and mineral spas

Helena Rodrigues, Ana Brochado and Michael Troilo

ABSTRACT

Thermal and mineral spa treatments are an important therapeutic tool backed by centuries of experience and numerous scientific studies showing its effectiveness. The research consisted of content analysis of 1,254 customer reviews posted online from 2015 to 2019. Qualitative analysis identified five key themes: thermal (spa), staff, room, location, and pool. On a theoretical level, this research's results help to fill research gaps in the literature by separating attributes linked to satisfaction from those related to dissatisfaction. This study is different as we focus on the factors affecting tourism satisfaction and dissatisfaction in thermal and mineral spas using sentiment analysis.

Introduction

The Global Wellness Institute (GWI) (2017) reports that the thermal spring market is expanding as tourists in a growing numbers are turning to water-based therapies for their health and wellness benefits. Most spa establishments (i.e. 95% of the world market) are concentrated in Asia and Europe, where facilities have experienced solid growth in both attendance and revenues (GWI, 2017). The numbers speak for themselves as this lucrative market's overall revenues increased by 5.2 billion United States dollars ($) from 2015 to 2017. In the same period, the number of establishments jumped from 27,507 in 109 countries to 34,057 in 127 countries employing 1.8 million people (GWI, 2017).

The extensive literature on thermal waters provides a comprehensive understanding of this sector based on multiple perspectives. Researchers have offered wide-ranging theoretical contributions in the areas of medicine, health, environmental sciences, and geothermal power resources. However, the existing research on tourism has yet to shed light on the specific issue of customer satisfaction with thermal and mineral spas even though both satisfaction and spas are topics studied in this field.

Previous studies have addressed satisfaction (Moghavvemi et al., 2017; Pyke, Hartwell, Blake, & Hemingway, 2016; Wang, 2012) or dissatisfaction (Zeelenberg & Pieters, 2004) in health and wellness tourism in general. Regarding customers' experiences, experts have had difficulty simultaneously identifying indicators of satisfaction with thermal spas and acknowledging, identifying, and managing dissatisfaction with these facilities. In addition, few studies of Europe's thermal and mineral spas have explored the wellness tourism experiences of different types of guests who express positive and negative reviews on the website Booking.com. The present research's has two main goals:

- First, to identify thermal and mineral spas attractiveness based on tourists' web reviews in a post experience phase.
- The second objective consists in analyzing the sentiments that express positive and negative emotions.

This paper is organized as follows. It examines a literature review of wellness tourism in mineral springs, including the spa benefits and wellness tourism satisfaction. The methodology section describes the sample profile and data analysis. The next two sections provide the results, along with a discussion of their implications for both academics and practitioners. Finally, this study's limitations and possible directions for future research are given in the conclusion.

Literature review

Tourism in mineral spring spas

In terms of wellness tourism, Europe has for centuries remained far in the lead, inventing a concept of truly holistic wellness that goes far beyond spas to include nutrition, traditional medicine, mindfulness, and a powerful connection to nature. Examples of this tourism sector's offer are baths, healing systems, hydrotherapies, diets, and mental health therapies (GWI, 2016). Wellness tourism is now being increasingly acknowledged as a crucial sector of tourism, both nationally and internationally (Hartwell, 2014).

Wellness tourism is, therefore, expected to continue expanding, and, since this trend shows no sign of disappearing, further growth in wellness tourism can be expected indefinitely (Voigt, Brown, & Howat, 2011). Overall, experts predict that the tourism industry will continue to have a quite strong impact on Europe's economy. The annual revenues from the spa facilities alone were, on average, $93.6 billion from 2015 to 2017.

Currently, the hotel and spa resort category includes the largest number of spas as more tourists turn to water for its healing properties (GWI, 2017). The most upscale hotels have expanded their services to attract tourists who now see spa therapy as the determinant ingredient in resort vacations rather than merely an extra amenity (Voigt et al., 2011). The increasing demands for these kinds of services could be a window of opportunity for potential investors in health and wellness tourism. According to Alonso-Álvarez (2012, p. 15):

> "What today is perhaps presumptuously described by the name of health and wellness tourism ... has three basic components[:] ... [m]ineral ... or [t]hermal [springs r]esorts, ... [s]pa facilit[ies] that do ... not depend on local natural mineral water sources, and thalassotherapy[. This] is an activity ... [that has seen] constant growth in recent decades in all the developed economies of Europe".

The present research focused on the first two components and wellness tourism, in particular, which is a holistic concept that includes tourists' desire to enhance their spiritual, psychological, physical, and social wellness (Bushell & Sheldon, 2009; Hartwell et al., 2016). Hobson and Dietrich (1995, p. 23) report that wellness tourism "is a mentally and physically healthy pursuit to follow in our leisure time." Notably, health tourism focuses primarily on the treatment of illnesses, while wellness tourism concentrates on prevention (Voigt et al., 2011). Academics and practitioners have paid particular note of the factors contributing to thermal and mineral spa competitiveness. The renewal of thermal spa tourism represent an escape from the beach and sun short stay vacation type of tourism the thermal spa tourism has the advantage of low seasonability and invite tourists to stay for long periods of relaxation focused on health prevention (Smith and Diekmann, 2017). Along with thermal treatments the modern spa play an important role in the tourism industry attracting family vacation to mineral bath pools (Deng, Kuo, & Chen, 2008; Radnic, Gracan, & Fister, 2009). In summary the Global Wellness Institute (2017) define that the attractiveness of thermal spa include swimming and bathing offers, recreational activities, lodging services and food and beverage.

SPA benefits for tourists

In 1948, the World Health Organization stated that "[h]ealth is a state of complete physical, mental, and social wellbeing and not merely the absence of disease and infirmity." After World War II, a shift occurred in the health needs of many populations, giving birth to a new wellness movement that reflected post-war advances in medicine, leading to fewer deadly diseases and lower mortality rates. Dunn (1959) first coined the term "wellness," including in this idea that physicians consider wellness to be a general condition of health encompassing wellbeing of the spirit, mind, and body. Many decades later, Mueller and Kaufman (2001) suggested that the main purpose of wellness services is to maintain or promote individuals' health.

One of the reasons for the growth of wellness tourism is the world's aging populations, especially the large numbers of baby boomers who are being increasingly targeted as an influential market segment by marketers. Many baby boomers are healthier, financially better off, and better educated than previous generations at that age, and baby boomers possess a greater desire to escape and seek authentic experiences (Patterson & Pegg, 2009). This generation also values treatments at medical thermal spas for their substantive health-related benefits, namely, preventative health treatments, nutrition and fitness, and health and wellness education (Patterson & Pegg, 2009).

In response to this demand, the thermal and mineral spring industry now includes structured and professionally delivered programs (Meyer, 2014) in the form of:

- Mineral spring spas – spa offerings with on-site natural mineral, thermal, or seawater resources, which are used in professionally administered hydrotherapy treatments
- European spa products focused on wellness and health
- Resort and/or hotel spas – most having sprung up or appearing in areas where spas are a profitable amenity added to traditional resort vacation opportunities

In order to develop more sustainable tourism, the European Community has co-sponsored the strengthening of cultural roots through which European and non-European tourists can discover how Europeans have lived since ancient times (Juul, 2015). The top destinations in this market include a large number of European countries with a long history of using thermal and mineral waters for curative and therapeutic treatments (Yeung & Johnston, 2017).

Thermal treatments are a set of therapeutic actions practiced in thermal spas, which are recommended or prescribed by family doctors or hydrologists (Constant, Guillemin, Collin, & Boulangé, 1998), including treatments based on:

- Hydrotherapy – baths, hot and cold shower baths, and water massages better known as balneotherapy
- Climatotherapy – favorable climatic conditions
- Heliotherapy – sunbathing
- Aerotherapy – air bathing
- Thalassotherapy – seawater and algae
- Aromatherapy – essential oils

Thermal spa treatments, therefore, are an important therapeutic tool supported by centuries of experience and numerous scientific studies that demonstrate its effectiveness (Maraver & Karagulle, 2012).

These treatments can be defined as a set of techniques that facilitate contact between the person to be treated (i.e. the spa user) and natural mineral water and other complementary therapies (Constant et al., 1998). In a more restricted sense, thermal spa treatments have a series of beneficial effects due to the specific composition of natural mineral water, complemented by the benefits derived from thermal spa environments and applications of relevant techniques (Edlin & Golanty, 2012). Treatments must be medically prescribed and carried out at a thermal spa facility. The effectiveness of hydrological medicine has been fully proven, and spa users benefit from the associated effects of the surrounding environment (Abertini, Teodori, & Conti, 2007).

The *Global Wellness Economy Monitor* (Yeung & Johnston, 2017) reports that over one-quarter of thermal and mineral springs establishments offer value-added spa services (e.g. massages and facials) along with bathing options. These facilities offer a wide range of services, so they bring in more revenue than those facilities with bathing only (Tsang & Qu, 2000; Yeung & Johnston, 2017).

Wellness tourism satisfaction

> *The wellness tourism is travel associated with the pursuit of maintaining or enhancing one's personal wellbeing* (GWI, 2016) p. 69. While satisfaction has been defined as *"the summary psychological state resulting when the emotion surrounding disconfirmed expectations is coupled with the consumer's prior feelings about the consumption experience"* (Oliver, 1981, p. 27).

In a post experience phase tourism can influence life satisfaction and make individuals happy (Gilbert & Abdullah, 2004; Nawijn, 2011). Satisfaction with one's health can predict life satisfaction, wellness tourism good experiences can positively affect individuals' health, family life, and relationships (Chen & Petrick, 2013; (Lončarić, Lončarić, & Marković, 2015). Individuals who experience emotional wellness are more prone to experience satisfaction and have a positive attitude towards life (Ryan & Deci, 2001).

Wellness tourists, in particular, have high expectations that must be satisfied, including self-realization and a sense of belonging (Meyer, 2014; Puczkó & Bachvarov, 2006). Compared with other tourists, these travelers are more successful in their professional and job careers and are particularly concerned about preventing physical problems and preserving their good health (Puczkó & Bachvarov, 2006). This type of tourist is driven by an enjoyment of luxurious experiences that can add value to their vacations through the gratification generated by – and satisfaction with – events (Kim, Chiang, & Tang, 2017). The provision of additional travel services can also lead to satisfaction in wellness tourism through, for example, training programs, physical fitness plans, and healthy diets (Kim et al., 2017).

The concept of satisfaction is similar across the extensive literature on this topic since most researchers agree that satisfaction reflects overall positive service evaluations and that it has significant implications for revisit intentions (Han & Hyun, 2015). Oliver (1980) points out that customer satisfaction is both an affective and cognitive reaction to satisfactory services and that customer dissatisfaction is the result of individuals comparing their real service quality experiences to their expectations.

Neal, Uysal, and Sirgy (2007) argue that, more specifically, satisfaction with leisure activities can have a positive effect on individuals' overall life satisfaction, so leisure and health are the main drivers behind tourists' trips to wellness and health destinations (Chen, Prebensen, & Huan, 2008; Ryff & Singer, 2008). Nawijn and Mitas (2012) also found that "wellness tourism influences different dimensions [and,] more specifically [,] the cognitive component of life satisfaction" (p. 533). Neal, Sirgy, and Uysal (1999); Sirgy (2010); and Sirgy, Kruger, Lee, and Yu (2010), in turn, concluded that tourists' satisfaction when traveling while on holiday can have a positive effect on their psychological wellness. The cited authors suggest that positive vacation experiences also affect how individuals evaluate multiple domains (e.g. work, family life, health, and recreation) and their overall life satisfaction.

Another important finding is that tourism incorporating wellness experiences enhances destinations' sustainability and promotes tourists' satisfaction. Buhalis (2000) argues that destination managers need to achieve a balance between sustainability in terms of local resources and all stakeholders. When this balance is achieved, host countries will benefit and tourists' satisfaction will increase.

Methodology

The first step when examining Web-generated content is to determine the most appropriate methodological approach. The main purpose of establishing a clear methodological approach is to ensure generalizable and reproducible results (Lai & To, 2015; Smith & Humphreys, 2006).

To satisfy the purposes of this study, the selected methods were systematically applied in four phases: 1) definition of goals and scope, 2) data collection, 3) sentiment classification, and 4) concept map analysis.

Data collection and information source

The data were collected from user-generated content (UGC) produced after tourists book accommodations on the Booking.com website (see www.booking.com). This website encourages tourists to post positive and negative reviews of their experiences, which means the reviews serve the present research's purposes. This UGC proved to be a useful source of data on tourists' experiences in thermal and mineral spa facilities (Lu & Stepchenkova, 2015). The present study was thus able to identify satisfaction and dissatisfaction attributes by separating the reviews into two groups by their satisfaction and dissatisfaction content, as well as generating specific concept maps and highlighting gender-related differences for each group (Cadotte & Tugeon, 1988).

Booking.com is the most popular travel-related site (The eBizMBA Rank Guide, 2019) with 40 million estimated unique monthly visitors, followed closely by TripAdvisor with 38 million (www.ebizmba.com).

For the present study, the spa facilities reviewed were segmented based on the information available from Termas de Portugal (see http://www.termasdeportugal.pt/). Ten hotels and/or resorts were selected from the 40 thermal and mineral spas in Portugal, based on reviews in Portuguese that included an associated rating of 8.0 or above out of 10 possible points and comments that have at least one complete sentence.

The sample consisted of 1,254 reviews with a total of 35,112 words. The retrieved reviews were those posted from January 2015 to January 2019. Additional information gathered from the Booking.com website included the hotels and/or resorts' name, number of stars, review scores, and ratings, as well as the reviewers' gender.

Data analysis with Leximancer

Leximancer 4.0 was used to address the first to identify factors contributing to thermal and mineral spa resorts' competitiveness. The software can be used to analyze quantitatively the relationships between textual. This inductive process helps researchers identify important "themes" by analyzing patterns and forming clusters based on relationships. Leximancer has previously been successfully used as a data analysis tool in marketing, hospitality, tourism (Brochado, Troilo, & Shah, 2017; Sainaghi, Phillips, & Zavarrone, 2017; Wu, Wall, & Pearce, 2014), health (Cretchley, Gallois, Chenery, & Smith, 2010), and accounting research (Crofts & Bisman, 2010). For each "keyword" identified, Leximancer finds the related concept, which facilitates deeper analyses and interpretations of the themes. Figure 1 presents an example of the operational method; Leximancer uses is a concept space approach in which words are mapped into a much smaller set of concepts (Chen et al., 1996).

Detailed information on Leximancer's algorithms has been widely described and discussed in the literature.

Sentiment analysis

To achieve the second research objective namely to identify the key drivers of wellness tourism satisfaction and dissatisfaction this study used the software Leximancer feature "Sentiment Lens" set automatically generate insight into positive and negative sentiments in the tourists' reviews. Sentiment Lens will only apply the sentiment terms that are identified as relevant and used consistently within all the reviews. This software's Automatic Sentiment Lens can thus identify words that

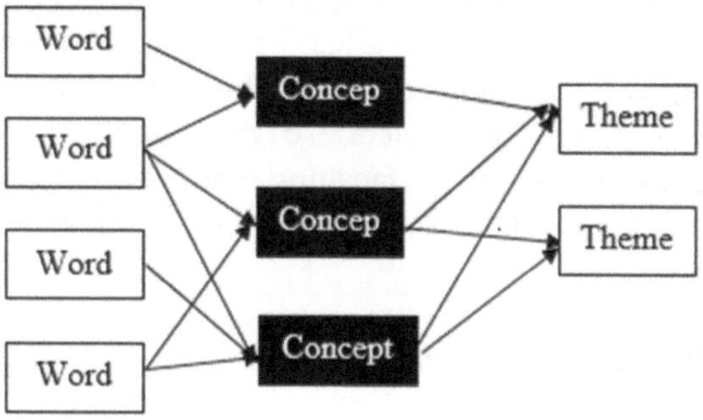

Figure 1. Semantic pattern model (Adapted from Leximancer, 2016).

reveal favorable sentiment terms and unfavorable or negative sentiment terms (Leximancer Pty Ltd, 2016). For example, in the phrase, "the service is unacceptable," is a negative description, sentiment analysis can provide further insights by organizing tourists' descriptions in terms of negative or positive appraisals (Pang & Lee, 2008). Sentiment analysis has also been described as attitude mining because it uses content analysis and computer language to capture meaningful information. This tool has often been used in marketing research and customer orientation studies to extract meaning from UGC (Schoen et al., 2013).

According to Turney (2002), sentiment analysis seeks to determine the attitude of speakers or writers toward some topic or the overall contextual polarity of documents. The strength of polarity is classified according to the content's closeness to extremely positive and extremely negative attitudes (e.g. "dirty," "good," "bad," "very," and "could not have done more to") (Turney, 2002).

Negative (i.e. weaknesses) and positive (i.e. strengths) reviews mention attributes that have been grouped into four categories in previous research focusing on hotel and restaurant hospitality (Cadotte & Tugeon, 1988). These attribute groups are satisfiers (i.e. positive attributes and/or strengths), dissatisfiers (i.e. negative attributes and/or weaknesses. Prior studies that have investigated positive and negative attitudes in customers' reviews on websites have further identified six categories: positive, negative, mixed, neutral, irrelevant, or uncertain (Godes & Mayzlin, 2004).

Results

Sample characteristics

The demographic information extracted from the reviews in the sample was relatively limited. Among the reviewers who reported their gender, more were male (59%) than female (41%). The majority of reviews (72%) included both positive and negative aspects related to their stay. Most facilities reviewed in the UGC collected are located in the north of Portugal (71%), and four star hotels were the most heavily represented (63%) in the reviews (see Table 1).

General description of reviews

To address the first aim of the present study, a concept map was generated to reveal the most common themes and concepts (see Figure 2) found in all 1,254 reviews (35,112 words). This map includes concepts – the smaller gray nodes – that are grouped into themes – the larger shaded circles (Leximancer Pty Ltd, 2016).

As can be seen in the figure above, the five dominant themes in post-experience customer reviews are "thermal" (1,474 hits), "staff" (1,260 hits), "room" (860 hits), "location" (492 hits), and "pool" (463 hits). A classification algorithm was used to conduct the text analysis to ensure reproducibility. Smith and Humphreys (2006) explain the process as follows:

> [Leximancer defines] blocks of *n* sentences and adds [up] their weightings in each concept. The software allows the extraction of [the] contextual information of text tags ... [in] negative and positive reviews and by male and female reviewers [. This] ... expand[s] the basic features of this research in terms of selecting the relevant concepts with likelihood percentages. (p. 268)

Table 1. Thermal mineral spa reviews.

Demographics	(N = 1254)	Percent
Reviews by Gender		
Female	505	41%
Male	749	59%
Total	*1254*	*100%*
Reviews by Type		
Only negative	150	12%
Only positive	201	16%
Negative and positive	903	72%
Total	*1254*	*100%*
Reviews by Geographic Location		
North	890	71%
Center	250	20%
South	114	9%
Total	*1254*	*100%*
Reviews by Hotel Star Rate		
3 Stars	326	26%
4 Stars	789	63%
5 Stars	139	11%
Total	*1254*	*100%*

Negative and positive reviews

For consumers, online reviews are now the most common way to register complaints, express feelings, comment on levels of satisfaction, and rate services or hotels (Yoo & Gretzel, 2008). For managers, online reviews represent the best channel through which to receive and assess additional information on service delivery and quality – especially negative reviews with low ratings because these are more likely to reflect substantive management problems (Schuckert, Liu, & Law, 2015). However, both negative and positive reviews were considered of interest in the present study as they can facilitate a fuller understanding of tourists' experiences during their stays at thermal and mineral spas.

Negative reviews

The negative reviews posted on Booking.com were identified by content analysis, which identified five dominant negative themes, as shown in Figure 3. The themes most closely associated with negative experiences are "service" (835 hits), followed by "thermal" (564 hits), "price" (276 hits), "restaurant" (102 hits), and "room" (78 hits). The "thermal (spa)" and "room (air-conditioner)" are the two major categories of complaints with high counts in reviews by female guests. The major categories of complaints expressed by male reviewers is "prices" and "exterior."

Positive reviews

Positive reviews of the wellness tourism facilities in question were identified by content analysis, which produced five positive themes, as shown in Figure 4. A concept map was generated with these themes, showing that the most prominent themes are "hotel" (945 hits), "thermal" (746 hits), "staff" (543 hits), "experience" (487 hits), and "relax" (67 hits). The categories of "pool" and "experience" were mentioned the most often in positive reviews by female guests, while major positive attributes identified by males are "restaurant" and "hotel."

Discussion

To identify thermal and mineral spas attractiveness based on tourists' web reviews in a post experience phase as mentioned previously, a general concept map was produced to identify the most common themes and concepts that represent thermal and mineral spa tourists' reviews (see Figure 2). The second aim was to examine how tourists express their negative (see Figure 3) or positive emotions (see Figure 4). The concept maps based on the negative and positive reviews indicate some common themes as in the General description of reviews based on the words frequency and co-occurrence (Smith & Humphreys, 2006). In this section we analyze the new themes that have emerged from the sentiment analyze.

General review evaluation

The UGC data used to identify the five main themes shown in Figure 2 above provide the basis for a solid understanding of each theme, as well as practical suggestions.

Thermal (and mineral spas)

The "thermal (spa)" facilities are the most frequently mentioned attribute. More than half of the customers who mentioned this attribute highlighted "The Thermal facilities excellent location very calming and the mineral water is worm ... it is the ideal place to rest ... ". One customer review said, " ... thermal treatments area is very pleasant and relaxing [. W]e returned home feeling rejuvenated."

These results confirm that wellness tourists value hydrotherapy and acknowledge that mineral and thermal waters have natural healing properties, as reported by Yeung and Johnston (2014). Thermal spa tourism in facilities close to natural mineral springs is similar to other European establishments in that they represent an escape from overcrowded capitals.

One characteristic specific to Portugal's thermal spa tourism serves a dual purpose. Medical hydrology is practiced within a clear legal framework and is highly regulated since only doctors can prescribe hydrotherapies, which are offered all year round and which can also diminish seasonality. The second purpose of thermal spa treatments is related to recreational activities. These specific services add value, and they can help reposition Portugal's spa resorts within thermal spa tourism in Europe and on other continents.

Staff

The "staff" was the second most frequently mentioned attribute. Reviews primarily highlight the staff's positive attitude. The complimentary comments made by guests include "the staff is super nice and attentive" and "[t]he staff is always ready to help you." The existing literature reveals that hotel staff is an important predictor of guests' level of trust. Trust comprises two aspects, namely, trust in employees or staff and trust in company practices (Sirdeshmukh, Singh, & Sabol, 2002). The first facet of trust relates to customers' perceptions of staff behavior and performance in service encounters. The second is mainly based on companies' performance (Ariffin & Maghzi, 2012).

Notably, the reviews examined in the present study often mention that the staff is unprepared, which can lead to service failures. The short-term success of spa hotels thus depends heavily on continually investing in proper staff training, reducing turnover, and cultivating employees' capabilities.

Room (quality)

The theme of "room" emphasizes rooms' quality as reflected in comments such as "[we had a q]uiet room, [which was] cozy [and] comfortable ... [and]

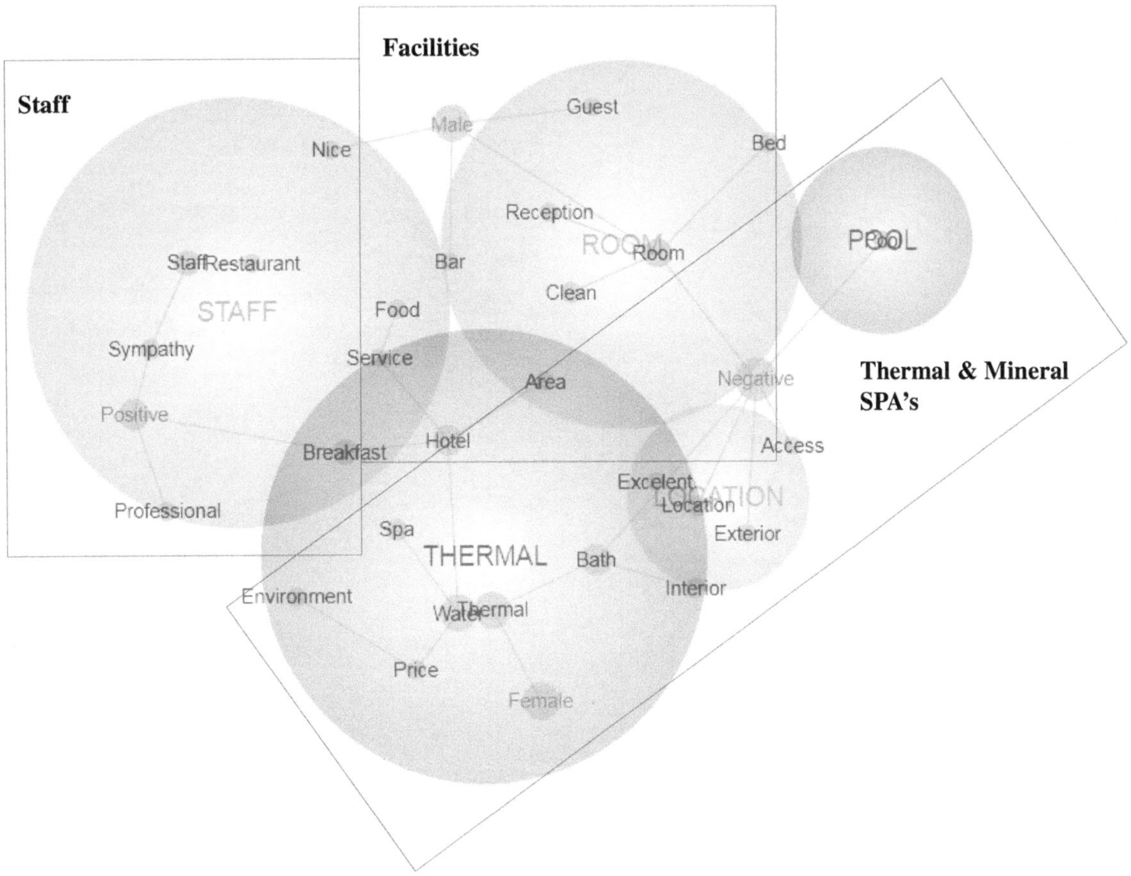

Figure 2. General concept map of reviews.

great for a few days of rest … . The rooms were … excellent and clean [with a] very comfortable bed." The GWI (2016) predicts that the future of European spas will include refocusing on nature and quiet environments, from the facilities' design to the guests' rooms.

Rooms that include clean private accommodations and a comfortable bed are strong predictors of guest satisfaction (Zhang & Cole, 2016). Zhang and Mao (2012) also concluded that well-maintained rooms that are clean and comfortable are important dimensions of customer satisfaction that, in turn, lead to a favorable hotel image.

The present results indicate that managers need to acknowledge wellness tourists' preferences when rebuilding – or designing new – facilities. Most hotels that offer wellness treatments focus on the design of the common areas (e.g. front desk and lounge) and tend to neglect or standardize the design and comfort of private accommodations. Thus, rooms should be improved to meet guests' expectations and encourage longer stays.

Location

"Location" attributes in reviews mainly highlight the easily "accessible location with magnificent views." One reviewer highlighted the "[c]lose location to other places of interest." Lu and Stepchenkova (2012) conducted a related study whose results also indicate that the rooms, nature, food, location, and value for money influence guests' satisfaction.

Geographical location and natural assets include the sun and sea in Portugal's western coastal and southern regions, as well as biodiversity, varied topography, and a rich historical heritage (Governo de Portugal, 2014). These assets can be more extensively leveraged by entrepreneurs, who need to develop thermal and mineral spa businesses that fuse unique settings, local traditions, and location-specific offerings with world-class services and built environments.

Pool

The theme of "pool" reflects how wellness tourism resorts use thermal mineral water pools for

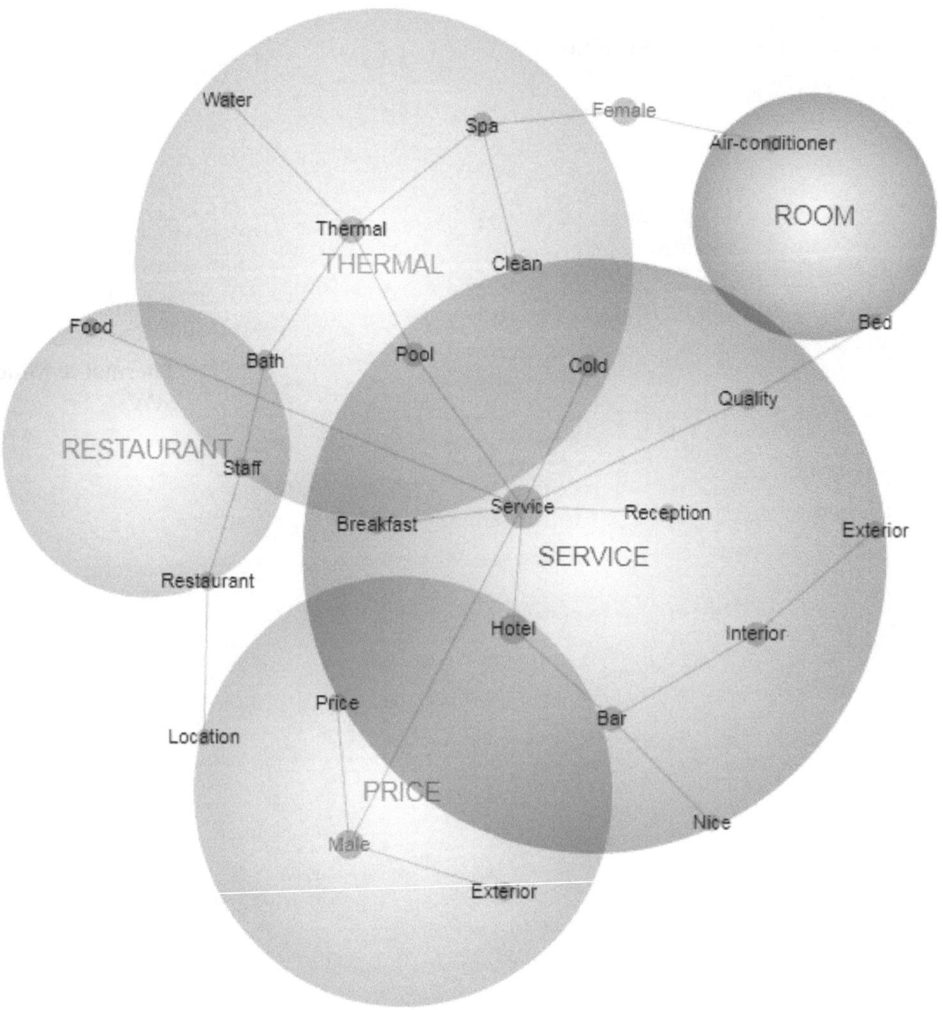

Figure 3. Negative reviews concept map.

treatments and recreational activities (GWI, 2017). However, pools are described in reviews as a primary source of dissatisfaction that ruins tourists' thermal and mineral spa experiences. For example, one review indicates that "the exterior access to the pool during cold months" should be improved and that "the water temperature" needs to be increased for comfort. Thermal pools should also include health services especially designed to promote spas' well-known therapeutic benefits, and pool areas need to be easily accessible. These results confirm Patterson and Pegg (2009) findings.

Negative reviews

Three new themes – "service", "price," and "restaurant" – emerged from the separate analysis of negative reviews. The most important negative attribute is "service" quality. One customer commented that "the food was very good, but the service was unprofessional." Another reviewer reported "old infrastructure and poor service." Overall, what differentiates one establishment from another is not only the food and spa facilities but also the quality of the services provided, which confirms Chu and Choi (2000) findings.

Service quality appraisal is particularly important for online communities as it gives potential guests indirect access to prior service experiences, on which prospective tourists can base their expectations of whether the companies in question will deliver quality services (Godes & Mayzlin, 2004). Service failures can thus be considered the main reason for customer dissatisfaction in spa resorts, and the resulting negative electronic word of mouth (e-WOM) needs to be closely monitored.

Unfavorable comments also include complaints about the "price" as compared to perceived service quality. One customer complained, "the price is too high for the quality of the service." Another guest wrote, "the rooms are very old fashioned, which makes the price too high." Since customer

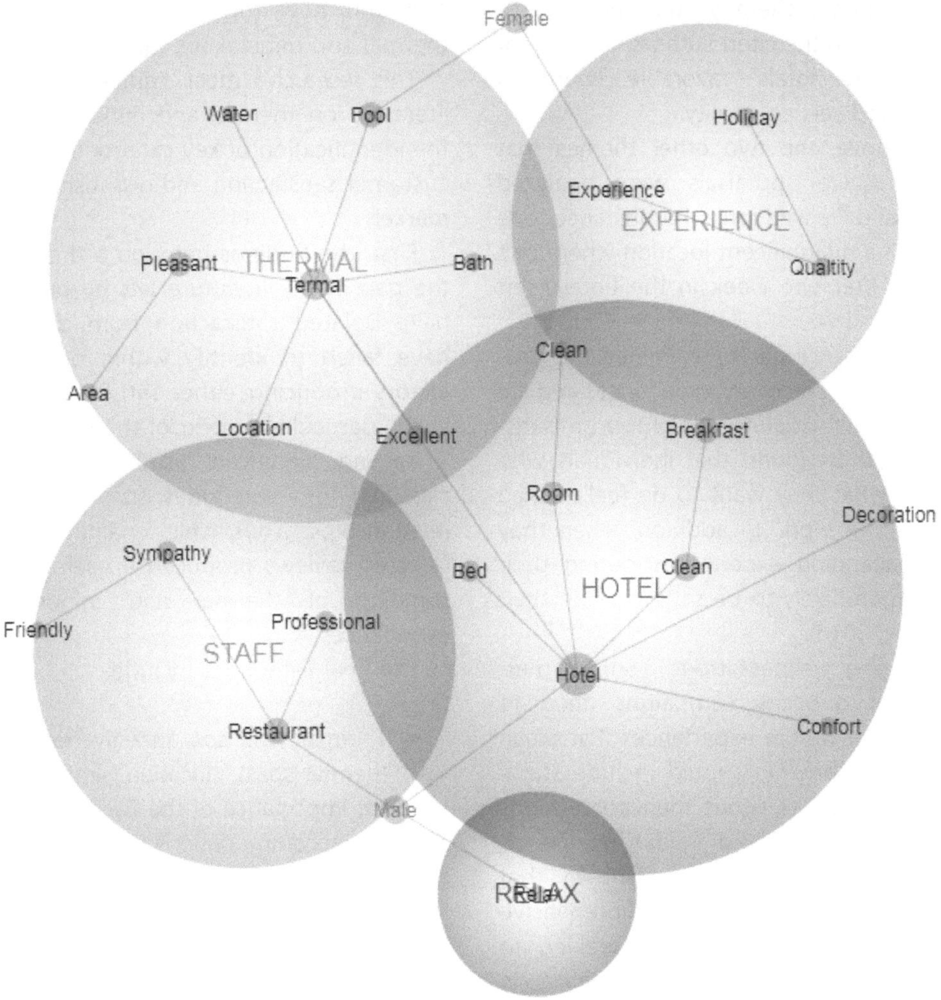

Figure 4. Positive reviews concept map.

satisfaction can reduce price sensitivity, increase customer loyalty, and increase cross-buying in the competitive wellness tourism market, spa resort managers need to foster more customer satisfaction and increase positive e-WOM (Deng & Pei, 2009).

The last negative theme is "restaurant," which is especially common among male guests. One reviewer particularly noted that "[t]he price, quality (and quantity) ratio of the restaurant is just not acceptable." Another guest commented on "[t]he lack [of] attention given by the staff [to clients] in the restaurant."

Customers evaluate not only food but also multiple aspects of service experiences, (Kim, Chen, & Jang, 2006). In spa hotels, the price customers pay includes the restaurant meals and services they receive, so, in this complex pricing environment, understanding guests' price perceptions and their role in customers' satisfaction is more effective than using objective pricing strategies (Han & Kim, 2009). More specifically, understanding the multiple aspects of services is particularly important in spas with full-service restaurants.

Positive reviews

Separate analysis extracted three new frequently mentioned themes – "hotel," "experience," and "relax(ation)" – from positive reviews. The results suggest that positive reviews strengthen hotels' overall image with statements such as "[this is a m]odern[,] comfortable hotel with an excellent view" or "[the f]antastic pool [was] very pleasant … [and] I enjoyed the gardens, the park, and the lake."

Various studies have found a clear relationship between positive perceptions of hotel brands and travelers' brand choices and positive WOM (Dimitrovski & Todorović, 2015). In addition, Kandampully and Hu (2007) underline that hotel image is influenced by both service quality and satisfaction and that these, in turn, enhance hotels' image, thereby strengthening customer loyalty. In line with the cited authors, the current study confirmed that hotels' favorable image is enhanced by improved service quality.

As mentioned above, the two other themes that appeared when positive appraisals were analyzed were "experience" and "relax(ation)." For instance, one guest wrote, "[this is a m]agnificent location where one feels relaxed … . [A]fter one week in the hotel I left feeling younger."

Chen, Huang, and Petrick (2016) examined the relationships between holiday recovery experiences and life satisfaction through the mediating role of tourism satisfaction. The cited authors found that individuals who are able to control what they want to do feel relaxed and disconnected from work. In addition, when they have new and challenging experiences during their holidays, they are more likely to be satisfied with their holidays (Chen et al., 2016).

Given the emerging wellness trend, tourism marketers in charge of publicity campaigns need to emphasize the relevant leisure experiences (Patterson & Pegg, 2009). Advertisements should include statements from wellness tourists about their perceptions of opportunities to escape and relax (Mueller & Kaufman, 2001). Market research has shown that most three- to five-star hotels provide comprehensive wellness facilities. Thus, wellness hotels seeking to increase their competitiveness must specialize in health information, individual care, and a wide range of cultural and relaxation offers.

Although the same hotels can host health and wellness tourists, these two segments have to be considered separately when selecting marketing strategies (Mueller & Kaufman, 2001). Examining spa resort reviewers' key positive and negative concepts offers critical information that can help managers identify realistic consumer service options. Their hotels can then develop appropriate strategies and gain competitive advantages in the wellness tourism sector.

Conclusion

The present research on thermal and mineral spa hotel businesses associated with wellness tourism included content analyses of key attributes of health facilities that offer therapeutic water treatments. Unlike many highly-structured assessments in previous satisfaction and valuation surveys, this study emphasized customers' natural language used in their depictions of spa experiences.

In addition, the qualitative methodology applied, while not pioneering, is still relatively new in the field, and, until now, this method has not been applied to thermal and mineral spa facilities.

This research's most significant contribution to the literature on thermal and mineral spa tourism is thus the identification of key categories of attributes driving customer satisfaction and dissatisfaction in this specific market.

First, the findings help, on a theoretical level, to fill the gap in the literature left by previous studies that have isolated satisfaction from dissatisfaction. These have failed to identify within one study the set of factors producing either satisfaction or dissatisfaction in the particular context of thermal and mineral spas.

Second, significant attributes were identified and classified into categories according to these factors' relationships. Thus, this qualitative analysis of the selected reviews produced a useful systematic understanding of thermal and mineral spas tourists' experiences.

The final categories include:

- "Thermal" and spa attractiveness – which includes "thermal [spas]," "location," and mineral water "pools"
- The importance of the Facilities – in particular, private accommodations (i.e. the "room" condition)
- Services provided by the "staff"

Tourists are not all the same. They have different images of their experiences, and their sociodemographics are heterogeneous (Dolnicar, 2008). Furthermore, the qualitative interpretations of the guest reviews examined in the present study further enhanced the accuracy of the above list of key thermal and mineral spas attractiveness.

Third, the results provide information on satisfaction and dissatisfaction in wellness tourism and highlight which favorable attributes can improve the quality of the wellness and health tourism products.

The results include practical advice for managers regarding the importance of thermal spa conditions and hotel facilities.

Previous research has indicated that the service quality of Portuguese hotels is still lagging behind international standards (Brochado, Rita, & Gameiro, 2015; Tsang & Qu, 2000). Therefore, an important way to leverage the demand for thermal and mineral spas and expand the wellness tourism sector is for entrepreneurs and local governments to develop more spas that fuse unique settings, local traditions, and place-specific offerings with world-class built environments and services.

Lodging businesses need to use these attributes to move from simply avoiding customer dissatisfaction to maximizing customer satisfaction through above-and-beyond service delivery (Zhang & Cole, 2016).

Additionally, results provide a strategic order of attributes based on customers' demographic attributes, which thermal and mineral spa facilities can use to implement or improve their service attributes and, thereby, maximize their guests' satisfaction. The main performance factor male customers perceive negatively is the overall "price." In contrast, the basic attributes that contribute to female clients' negative perceptions are the "thermal (spa)" and "room (air-conditioner)." These factors should, therefore, be given priority overall when examining guests' evaluations of hotels. To be successful, managerial strategies need to be able to increase tourism expenditure and visitor numbers by providing satisfying, memorable experiences (Pyke et al., 2016).

Notably, this study's results rely on the assumption that all the reviews were posted by guests who actually stayed in the facilities reviewed (Ayeh, Au, & Law, 2013). In terms of future research, a question worth investigating is how Internet usage in health and wellness trip planning differs between tourists who seek leisure experiences and those who travel for mineral water treatments. Another recommended line of research would be to focus on additional factors that were not included in this research but that could also contribute to a better understanding of tourists' choice of health and wellness tourism service providers (Medina-Muñoz & Medina-Muñoz, 2014).

Disclosure statement

No potential conflict of interest was reported by the authors.

ORCID

Helena Rodrigues http://orcid.org/0000-0001-7621-824X
Ana Brochado http://orcid.org/0000-0002-8917-2575

References

Abertini, M. C., Teodori, L., & Conti, M. E. (2007). Drinking mineral waters: Biochemical effects and health implications—The state-of-the-art. *International Journal Environmental Health*, *1*(1), 153–169.

Alonso-Álvarez, L. (2012). The value of water: The origins and expansion of thermal tourism in Spain, 1750–2010. *Journal of Tourism History*, *4*(1), 15–34.

Ariffin, A. A. M., & Maghzi, A. (2012). A preliminary study on customer expectations of hotel hospitality: Influences of personal and hotel factors. *International Journal of Hospitality Management*, *31*(1), 191–198.

Ayeh, J. K., Au, N., & Law, R. (2013). "Do we believe in TripAdvisor?" Examining credibility perceptions and online travelers' attitude toward using user-generated content. *Journal of Travel Research*, *52*(4), 437–452.

Brochado, A., Rita, P., & Gameiro, C. (2015). Exploring backpackers' perceptions of the hostel service quality. *International Journal of Contemporary Hospitality Management*, *27*(8), 1839–1855.

Brochado, A., Troilo, M., & Shah, A. (2017). Airbnb customer experience: Evidence of convergence across three countries. *Annals of Tourism Research*, 160–162. doi:10.1016/j.annals.2017.01.001

Buhalis, D. (2000). Marketing the competitive destination of the future. *Tourism Management*, *21*(1), 97–116.

Bushell, R., & Sheldon, P. J. (Eds.). (2009). *Wellness and tourism: Mind, body, spirit, place*. New York: Cognizant Communication Corporation.

Cadotte, E. R., & Tugeon, N. (1988). Dissatisfiers and satisfiers: Suggestions from consumer complaints and compliments. *Journal of Consumer Satisfaction, Dissatisfaction and Complaining Behavior*, *1*(1), 74–79.

Chen, C. C., Huang, W. J., & Petrick, J. F. (2016). Holiday recovery experiences, tourism satisfaction and life satisfaction - Is there a relationship? *Tourism Management*, *53*, 140–147.

Chen, C. C., & Petrick, J. F. (2013). Health and wellness benefits of travel experiences: A literature review. *Journal of Travel Research*, *52*(6), 709–719.

Chen, H., Schatz, B., Ng, T., Martinez, J., Kirchhoff, A., & Chienting, L. (1996). A parallel computing approach to creating engineering concept spaces for semantic retrieval: The Illinois digital library initiative project. *IEEE Transactions on Pattern Analysis and Machine Intelligence*, *18*(8), 771–782.

Chen, J. S., Prebensen, N., & Huan, T. C. (2008). Determining the motivation of wellness travelers. *Anatolia*, *19*(1), 103–115.

Chu, R. K. S., & Choi, T. (2000). An importance-performance analysis of hotel selection factors in the Hong Kong hotel industry: A comparison of business and leisure travellers. *Tourism Management*, *21*(4), 363–377.

Constant, F., Guillemin, F., Collin, J. F., & Boulangé, M. (1998). Use of SPA therapy to improve the quality of life of chronic low back pain patients. *Medical Care*, *36*(9), 1309–1314.

Cretchley, J., Gallois, C., Chenery, H., & Smith, A. (2010). Conversations between carers and people with schizophrenia: A qualitative analysis using Leximancer. *Qualitative Health Research*, *20*(12), 1611–1628.

Crofts, K., & Bisman, J. (2010). Interrogating accountability: An illustration of the use of Leximancer software for qualitative data analysis. *Qualitative Research in Accounting & Management*, *7*(2), 180–207.

Deng, W., Kuo, Y., & Chen, W. (2008). Revised importance performance analysis: Three-factor theory and benchmarking. *The Service Industries Journal*, *28*(1), 37e 51.

Deng, W. J., & Pei, W. (2009). Fuzzy neural based importance-performance analysis for determining critical service attributes. *Expert Systems with Applications*, *36*(2 PART 2), 3774–3784.

Dimitrovski, D., & Todorović, A. (2015). Clustering wellness tourists in spa environment. *Tourism Management Perspectives*, *16*, 259–265.

Dolnicar, S. (2008). Market segmentation in tourism. In D. Woodside & A. G. Martin (Eds.), *Tourism management: Analysis, behaviour and strategy* (pp. 129–150). Wallingford, UK: CABI Books.

Dunn, H. L. (1959). High-level wellness for man and society. *American Journal of Public Health*, *49*(6), 786–792.

Edlin, G., & Golanty, E. (2012). *Health and wellness*. Burlington, MA: Jones & Bartlett Publishers.

Gilbert, D., & Abdullah, J. (2004). Holidaytaking and the sense of well-being. *Annals of Tourism Research*, *31*(1), 103–121.

Godes, D., & Mayzlin, D. (2004). Using online conversations to study word-of-mouth communication. *Marketing Science*, *23*(4), 545–560.

Governo de Portugal. (2014). Relatório do grupo de trabalho interministerial: Turismo de saúde [Report of inter-ministerial work group: Health tourism]. Lisbon: Governo de Portugal.http://healthportugal.com/Quem%20somos/documentos/relatorio-do-grupo-de-trabalho-interministerial.pdf

Governo de Portugal. (2014). *Relatório do grupo de trabalho interministerial: Turismo de saúde* [Report of interministerial work group: Health tourism]. Lisbon: Author.

GWI. (2016). *Ten predictions for the future of wellness, travel, spa and beauty in Europe*. Retrieved from https://www.globalwellnesssummit.com/press/press-releases/press-release-ten-predictions-for-the-future-of-wellness-travel-spa-and-beauty-in-europe/

GWI. 2017 Global wellness economy monitor. Retrieved from https://globalwellnessinstitute.org/industry-research/global-wellness-economy-monitor-2017/

Han, H., & Hyun, S. S. (2015). Customer retention in the medical tourism industry: Impact of quality, satisfaction, trust, and price reasonableness. *Tourism Management*, *46* (February), 20–29.

Han, H., & Kim, W. (2009). Outcomes of relational benefits: Restaurant customers' perspective. *Journal of Travel & Tourism Marketing*, *26*(8), 820–835.

Hartwell, H. (2014). Wellness tourism: A destination perspective. *Tourism Management*, *42*, 305–306.

Hartwell, H., Fyall, A., Willis, C., Page, S., Ladkin, A., & Hemingway, A. (2018). Progress in tourism and destination wellbeing research. *Current Issues in Tourism*, *21*(16), 1830-1892.

Hobson, J. S. P., & Dietrich, U. C. (1995). Tourism, health and quality of life. *Journal of Travel & Tourism Marketing*, *3*(4), 21–38.

Juul, M. (2015). *Tourism and the European community: Recent trends and policy developments*. Retrieved from http://www.europarl.europa.eu/thinktank/en/document.html?reference=EPRS_IDA(2015)568343

Kandampully, J., & Hu, -H.-H. (2007). Do hoteliers need to manage image to retain loyal customers? *International Journal of Contemporary Hospitality Management*, *19*(6), 435–443.

Kim, E., Chiang, L., & Tang, L. (2017). Investigating wellness tourists' motivation, engagement, and loyalty: In search of the missing link. *Journal of Travel and Tourism Marketing*, *34* (7), 867–879.

Kim, H. J., Chen, M.-H., & Jang, S. (2006). Tourism expansion and economic development: The case of Taiwan. *Tourism Management*, *27*(5), 925–933.

Lai, L. S. L., & To, W. T. (2015). Content analysis of social media: A grounded theory approach. *Journal of Electronic Commerce Research*, *16*(2), 138–152. Retrieved from http://www.jecr.org/sites/default/files/16_2_p05.pdf

Leximancer Pty Ltd. (2016). *Leximancer user guide, release 4.5* [PDF document]. Retrieved from https://doc.leximancer.com/doc/LeximancerManual.pdf

Lončarić, D., Lončarić, D., & Marković, S. (2015). Health tourism, customer satisfaction and quality of life : The role of specialty hospitals. *ToSEE – Tourism in Southern and Eastern Europe: Sustainable Tourism, Economic Development and Quality of Life*, *3*, 159–173.

Lu, W., & Stepchenkova, S. (2012). Ecotourism experiences reported online: Classification of satisfaction attributes. *Tourism Management*, *33*(3), 702–712.

Lu, W., & Stepchenkova, S. (2015). User-generated content as a research mode in tourism and hospitality applications: Topics, methods, and software. *Journal of Hospitality Marketing & Management*, *24*(2), 119–154.

Maraver, F., & Karagulle, M. Z. (2012). Medical hydrology and balneology: Environmental aspects. *Monografias, Balnea-Series De*, *6*, 1–412.

Medina-Muñoz, D. R., & Medina-Muñoz, R. D. (2013). The attractiveness of wellness destinations: An importance-performance-satisfaction approach. *International Journal of Tourism Research*, *16*, 521–533.

Medina-Muñoz, D. R., & Medina-Muñoz, R. D. (2014). The attractiveness of wellness destinations: An importance-performance-satisfaction approach. *International Journal of Tourism Research*, *16*(6), 521–533.

Meyer, B. (2014). Health resort qualities as basis of brand tourism products. *Tourism Role in the Regional Economy*, *3*, 1-493.

Moghavvemi, S., Ormond, M., Musa, G., Isa, C. R. M., Thirumoorthi, T., Mustapha, M. Z. B., & Chandy, J. J. C. (2017). Connecting with prospective medical tourists online: A cross-sectional analysis of private hospital websites promoting medical tourism in India, Malaysia and Thailand. *Tourism Management*, *58*, 154–163.

Mueller, H., & Kaufman, E. L. (2001). Wellness tourism: Market analysis of a special health tourism segment and implications for the hotel industry. *Journal of Vacation Marketing*, *7* (1), 5–17.

Nawijn, J. (2011). Determinants of daily happiness on vacation. *Journal of Travel Research*, *50*(5), 559–566.

Nawijn, J., & Mitas, O. (2012). Resident attitudes to tourism and their effect on subjective wellbeing: The case of Palma de Mallorca. *Journal of Travel Research*, *51*, 531–541.

Neal, J. D., Sirgy, M. J., & Uysal, M. (1999). The role of satisfaction with leisure travel/tourism services and experience in satisfaction with leisure life and overall life. *Journal of Business Research*, *44*(3), 153–163.

Neal, J. D., Uysal, M., & Sirgy, J. M. (2007, November 1). The effect of tourism services on travelers' quality of life. *Journal of Travel Research*, *46*, 154–163.

Oliver, R. L. (1980). A cognitive model of the antecedents and consequences of satisfaction decisions. *Journal of Marketing Research*, *17*, 46–69.

Oliver, R. L. (1981). Measurement and evaluation of satis faction process in a retail setting. *Journal of Retailing*, *57*(3), 25–48.

Pang, B., & Lee, L. (2008). Opinion mining and sentiment analysis. *Foundations and Trends® in Information Retrieval*, *2*(1–2), 1–135.

Patterson, I., & Pegg, S. (2009). Marketing the leisure experience to baby boomers and older tourists. *Journal of Hospitality Marketing & Management*, *18*(2–3), 254–272.

Puczkó, L., & Bachvarov, M. (2006). Spa, bath, thermae: What's behind the labels? *Tourism Recreation Research, 31*(1), 83–91.

Pyke, S., Hartwell, H., Blake, A., & Hemingway, A. (2016). Exploring well-being as a tourism product resource. *Tourism Management, 55*, 94–105.

Radnic, R., Gracan, D., & Fister, M. (2009). Repositioning of thermal spa tourism of North-west Croatia in accordance with the European thermal spa tourism trends. *International Journal Of Multidisciplinary Research For South And South-Eastern Europe, 15*(1), 73–84.

Ryan, R. M., & Deci, E. L. (2001). On hapiness and human potentials: A review of research on hedonic and eudaimonic Well-being. *Annual Review Psychology, 52*, 141–166.

Ryff, C. D., & Singer, B. H. (2008). Know thyself and become what you are: A eudemonic approach to psychological well-being. *Journal of Happiness Studies, 9*, 13–39.

Sainaghi, R., Phillips, P., & Zavarrone, E. (2017). Performance measurement in tourism firms: A content analytical meta-approach. *Tourism Management, 59*, 36–56.

Schoen, H., Gayo-Avello, D., Metaxas, P. T., Mustafaraj, E., Strohmaier, M., & Gloor, P. (2013). The power of prediction with social media. *Internet Research, 23*(5), 528–543.

Schuckert, M., Liu, X., & Law, R. (2015). Hospitality and tourism online reviews: Recent trends and future directions. *Journal of Travel & Tourism Marketing, 32*(5), 608–621.

Sirdeshmukh, D., Singh, J., & Sabol, B. (2002). Consumer trust, value, and loyalty in relational exchange. *Journal of Marketing, 66*(1), 15–37.

Sirgy, M. J. (2010). Toward a quality-of-life theory of leisure travel satisfaction. *Journal of Travel Research, 49*(2), 246–260.

Sirgy, M. J., Kruger, P. S., Lee, D.-J., & Yu, G. B. (2010, March 19). How does a travel trip affect tourists' life satisfaction? *Journal of Travel Research*. doi:10.1177/0047287510362784

Smith, A. E., & Humphreys, M. S. (2006). Evaluation of unsupervised semantic mapping of natural language with Leximancer concept mapping. *Behavior Research Methods, 38*(2), 262–279.

Smith, M. K., & Diekmann, A. (2017). Tourism and wellbeing. *Annals Of Tourism Research, 66*, 1–13. doi:10.1016/j.annals.2017.05.006

Tsang, N., & Qu, H. (2000). Service quality in China's hotel industry: A perspective from tourists and hotel managers. *International Journal of Contemporary Hospitality Management, 12*(5), 316–326.

Turney, P. (2002, July). Thumbs up or thumbs down? semantic orientation applied to unsupervised classification of reviews. *In Proceedings of the 40th Annual Meeting on Association for Computer Linguistics, 8–10*, 417–424.

Voigt, C., Brown, G., & Howat, G. (2011). Wellness tourists: In search of transformation. *Tourism Review, 66*(1), 16–30.

Wang, H.-Y. (2012). Value as a medical tourism driver. *Managing Service Quality, 22*(5), 465–491.

Wu, M. Y., Wall, G., & Pearce, P. L. (2014). Shopping experiences: International tourists in Beijing's Silk Market. *Tourism Management, 41*, 96–106.

Yeung, O., & Johnston, K. (2014). *Global spa and wellness economy monitor*. Miami, Fl: Global Wellness Institute.

Yeung, O., & Johnston, K. (2017). *Global spa and wellness economy monitor*. Miami, Fl: Global Wellness Institute.

Yoo, K. H., & Gretzel, U. (2008). What motivates consumers to write online travel reviews? *Information Technology & Tourism, 10*(4), 283–295.

Zeelenberg, M., & Pieters, R. (2004). Beyond valence in customer dissatisfaction: A review and new findings on behavioral responses to regret and disappointment in failed services. *Journal of Business Research, 57*(4), 445–455.

Zhang, J. J., & Mao, Z. (2012). Image of all hotel scales on travel blogs: Its impact on customer loyalty. *Journal of Hospitality Marketing & Management, 21*(2), 113–131.

Zhang, Y., & Cole, S. T. (2016). Dimensions of lodging guest satisfaction among guests with mobility challenges: A mixed-method analysis of web-based texts. *Tourism Management, 53*, 13–27.

Index

Note: Page numbers in *italics* refer to figures